A New Paradigm for the Conquest of Land by Verteb
Exaptations

A New Paradigm for the Conquest of Land by Vertebrates That Includes Exaptations

By

Mauro Luís Triques
Martin Lindsey Christoffersen

BRILL

LEIDEN | BOSTON

Cover illustration: Coelacanth model, Devonian (Daderot, 2012) (exhibit in the Houston Museum of Natural Science, Houston, Texas, USA; photography was permitted in the museum without restriction).

This book is a revised and expanded English translation from the original Portuguese edition *Peixes conquistam a terra firme: nova abordagem para um evento acidental único*, published by Editora Baraúna in 2017.

Library of Congress Cataloging-in-Publication Data

Names: Triques, Mauro Luís, author. | Christoffersen, Martin Lindsey, author.
Title: A new paradigm for the conquest of land by vertebrates that includes exaptations / by Mauro Luís Triques, Martin Lindsey Christoffersen.
Other titles: Peixes conquistam a terra firme. English
Description: Boston, Massachusetts : Brill, [2024] | This book is a revised and expanded English translation from the original Portuguese edition Peixes conquistam a terra firme: nova abordagem para um evento acidental único, published by Editora Baraúna in 2017. | Includes bibliographical references and index.
Identifiers: LCCN 2023046473 (print) | LCCN 2023046474 (ebook) | ISBN 9789004684751 (hardback) | ISBN 9789004684911 (ebook)
Subjects: LCSH: Vertebrates–Evolution. | Vertebrates–Adaptation.
Classification: LCC QL607.5 .T7513 2024 (print) | LCC QL607.5 (ebook) | DDC 596–dc23/eng/20231101
LC record available at https://lccn.loc.gov/2023046473
LC ebook record available at https://lccn.loc.gov/2023046474

Typeface for the Latin, Greek, and Cyrillic scripts: "Brill". See and download: brill.com/brill-typeface.

ISBN 978-90-04-68475-1 (hardback)
ISBN 978-90-04-68491-1 (e-book)
DOI 10.1163/9789004684911

This book is printed on acid-free paper and produced in a sustainable manner.

To
Marcos Rodrigues
and
Alfredo Hannemann Wieloch,
colleagues and friends

Marcos brought to us the idea of producing a book on our unorthodox ideas and persuaded us into action in moments of wavering confidence

Alfredo suggested that the first author should interact with the second at the time of a preliminary manuscript of our first paper on the matter in the now long past year of 2007, which ultimately led to this book

Both provided the necessary support for our ideas under successive rejections by referees and editors!

∵

Contents

Preface IX
Acknowledgements XIII
List of Illustrations XVII
Notes for Reading This book XIX

1 **Phylogenetic Systematics or Cladistics, Introductory Information** 1
1 From Darwin to Hennig 1
2 Which Animals Are We Dealing With? 4
3 The Skeleton 11

2 **Adaptive Scenarios** 20

3 **Adaptationism, Microevolution, Macroevolution, and Exaptation** 33

4 **Monophyly and Geological Time as Counter-arguments to the Adaptationist View of the Origin of Terrestrial Vertebrates** 44
1 Monophyly of Terrestrial Vertebrates 44
2 The Geologic Time Gap 49
3 The Extreme Diversity of the Actinopterygii 52
4 The Phyletic Nature of †Placoderms and †Acanthodians 60

5 **Method Based on Cladistics: New Perspectives** 68
1 Criticisms of Adaptationism 79
2 Characters 81
3 Life Styles 102
4 The Issue of Topology 107
5 The Origins of Limbs with Digits, Walking, and Terrestriality 114
6 Recent Phylogenies of Basal Tetrapoda Reveal an Additional Argument 117
7 Exaptation: Misunderstandings of the Concept 123
8 Complementary Considerations 134

6 **Geological Time, Morphology, and Ecology** 135
1 Characters 140
2 Abrupt Appearance of Several Characters 144
3 Evolutionary Space for Independent New Invasions of Land by Vertebrates after the Devonian 146

7 **Absence of Correlation among Characters and Asynchrony in Their Origins as Additional Arguments** 148

8 **Synergy among Arguments and Summary Considerations of the Main Arguments** 152

9 **Exaptations, Not Preadaptations, at the Origin of Terrestriality in Vertebrates** 163

10 **Science and Its Limits** 192
- 1 A Statement on Scientific Paradigms 192
- 2 A View on the Sociology of Science 193
- 3 Epilogue 195

Glossary 197
References 208
Index 218

Preface

A New Paradigm for the Conquest of Land by Vertebrates That Includes Exaptations represents a new paradigm for studying and understanding biological evolution. This book criticizes the prevailing view, the adaptationist, microevolutionary outlook, as being the exclusive or main evolutionary process responsible for vertebrates to have occupied the terrestrial environment. We argue that macroevolutionary processes are significantly more important to explain an improbable evolutionary event. We conclude that macroevolutionary processes are the dominant factors involved in the origin of terrestriality. This is the view that emerges from our research, and which we will detail in the subsequent Chapters. Only independent researchers, without commitments to project approvals, restraints that usually accompany financial support, and with the encouragement of editors seeking new ideas in evolutionary biology, would be able to point out and argue for a different perception in the field.

Since Charles Darwin and Alfred Wallace (1858) proposed the theory of natural selection of organisms, as a mechanism generating new species, it became clear that the environment selects variations found among organisms. In current terms, this process of adaptation favors the differential passage of DNA that supports these variations in future generations. Because the environment changes through time, as becomes evident from geological studies, it is to be expected that organisms will be correspondingly affected. Thus, great importance has been given to the knowledge and comprehension of the environmental conditions along geological time. A great effort has been directed towards comprehending how this gradually changing environment acted for natural selection of organisms to occur and how new forms of life have been generated. This also applies to tetrapods, which are the vertebrates that made the transition from water to land. The environment changes continuously along geological time. Variations in aquatic to terrestrial forms were also interpreted as occurring continuously, resulting in progressive increases towards terrestriality. Changes would have occurred slowly, pace-like, greatly as a function of geological variations. However, even in textbooks (for example, that of Futuyma, 1995, page 20), it is recognized that not all evolution consists of adaptations resulting from natural selection. There are several selective factors, beyond those supplied by the ecological context, such as internal relations among biochemical paths and alternative ontogenetic paths, among others. Yet it becomes clear, from reading the specialized and detailed books on the subject of the origin of terrestriality in Tetrapoda (for example, Clack 2002, 2012; Laurin 2010; McGhee 2013), that the focus on the environment was maintained in the forefront in the mind of researchers specializing on this theme. To be fair, Laurin (2010: 88, 135, 138) already accepts exaptation as a punctual process for the water-land transition in Vertebrata. In our view, however, this process becomes the main process in evolution and represents the core concept of this book. The research strategy that identifies adaptations and the specific selective forces that directed evolutionary changes in past environments is termed adaptationism (Gould and Lewontin 1979). On the other hand, the effect of evolutionary chance was indicated with clarity by Gould and Vrba (1982) and developed in several ways. They are not referring to the chance event of DNA variation, but to aleatory, the chance event of a feature that emerged as a result of natural selection of DNA variation for one use. Subsequently such an original use could also fortuitously become useful for another, different use, called an effect. Clearly these features are adaptations for the first use, but not for the second use. For this reason, they were instead called exaptations. A position critical to adaptationism was clearly debated by Williams (1966) and Gould and Lewontin (1979), but has not been further explored in depth. This book presents an alternative view to the current perception that geological and ecological changes along the Devonian period generated environmental selective pressures on vertebrate life forms that led vertebrates out of the water and/or towards land, in a step-by-step process that can be called microevolution. We show in this book that independent features were selected for use in water, but collectively (and only collectively) they happened to be also useful on land, in a unique chance event called macroevolution.

We argue mainly on the basis of a cladistic framework. As far as we are aware, we are amongst the first authors to do so on the theme, following McLennan (2008). Because she preferred to use the term co-option, we were slow in locating this particular article. Possibly we are amongst the pioneers using this outlook also in a general way. We only depended on logic provided by the human brain to uncover a new way of interpreting the conquest of land by vertebrates. We do not rely on new technologies and did not require new data. Therefore, we hope to demonstrate to the reader that reinterpretation is always possible. The brain should be left free of prejudices (including the scientific ones), and of mental straightjackets of any kind.

This is thus not a book on information, but a book on ideas. We do not intend to bring new information or quantity of information to the reader, or even to present information in a didactic way. We hope to bring a series of new ideas on evolutionary biology that are in conflict with previous thought. These new ideas and propositions are not being made available to the general public, students, biologists, and not even to evolutionary biologists in general. This process is clearly shown by Gould and Lewontin (1979), Andrews *et al.* (2002), among many others. An overall account on conflicts of ideas in Science may be found in Charlton (2008). Indeed, the above cited authors picture these conflicts almost as a silent "war" for the control of our thinking. This "war" may be illustrated by Gould and Vrba (1982: 11) having pointed out that the arrangement of bones and muscles of tetrapods were not selected for terrestrial life, being instead exaptations for life on land. Yet the paleontologists specialized in the theme do not mention this. Further, this kind of evolutionary possibility is not mentioned in textbooks, nor even in evolutionary textbooks. The emergence of a feature for one use and its later different uses in the phylogeny of correlated biological entities in other ways, is not emphasized. Students and new specialists simply are unaware of these alternative thoughts in biological evolution. This book represents a "battle" in this war. Paleontologists working on the passage of vertebrates from water to land are strictly adaptationists: only natural selection is considered as producing step-by-step changes that finally permitted vertebrates to live on land. We want people to learn that there are alternatives. We have contrary arguments: there are other mechanisms involved in this subject. By means of this book, we would like to uncover what has been concealed from the eyes of the reader: new ideas on evolutionary biology have not been included in textbooks, popular Science books, popular Science magazines, and specialized publications. Evolutionary biology is much more than DNA variation (as mutations and the like) combined with the action of natural selection. A great deal of the evolutionary events occurring in the past were of a different kind, of which even specialists have not been aware of. We can trace this "war" back to at least Gould and Lewontin's (1979) seminal paper, that has been "forgotten" by several evolutionary biologists throughout the World, including paleontologists. Gould and Lewontin predicted that the new ideas would be simply put aside and forgotten. We are resurrecting their fight, as much as possible for the present and future generations. Many others between them and us have been fighting this "war", and we are bringing their arguments to the core of this book: the passage of vertebrates from water to land in an evolutionary perspective.

We hope to show the reader a new and different way to deal with each and all the set of biological information in an evolutionary perspective. We expect to convince the reader of the power of phylogenetic relationships expressed in a branched diagram (in this case, a cladogram) that is at the same time "deep" and "wide". This perspective may open your mind for a new set of thinking possibilities in evolutionary biology.

Our aim, as well as influencing the specialist on the theme, is to dialogue with any biologist interested in evolution. Readers must master the subject of phylogenetic systematics or cladistics. The glossary will help readers less acquainted with the theme.

We would like to make it clear that this work is not intended as a criticism of natural selection as an important evolutionary process. Natural selection generates biological diversity and adaptation of characters to the environment. What we criticize is viewing natural selection as the exclusive or main evolutionary mechanism, when there are other, already known, additional processes. We argue that these alternative evolutionary processes were fundamental for the passage of vertebrates from water to land. Probably these macroevolutionary processes will be demonstrated to be important also for other evolutionary transitions, particularly for evolutionary events of a fortuitous and highly improbable nature.

We hope the reader will enjoy a pleasurable trip along geological time. That he/she will be able to verify changes in the characters of these animal subjects. We hope the reader may become enthusiastic with these very interesting beings, those vertebrates and their ancestors that invaded land. We hope that the interested reader will find more than adaptationist explanations to a changing environment. And that his/her reasoning may evolve such as ours. Additionally, we hope the reader concludes with us that not only chance, but also luck, are events affecting our most remote biological origins, as well as our daily lives.

We will appreciate any comments or corrections (send to maurotriques@yahoo.com.br or martinlc.ufpb@gmail.com).

Acknowledgements

It is with enormous satisfaction that Mauro Luís Triques and Martin Lindsey Christoffersen acknowledge those persons that with great good will have offered their private photos in order to enable us to produce the Portuguese text, and first edition of this book. Many of these people are not, or were not, at the time, personally known to both authors: Olga Ferrarin, Lucas Menezes Silva (The Scientific Ink), Rafael de Castro Teixeira, and Paulo Roberto Petersen Hofmann (Biotemas magazine). José Cassimiro, a former undergraduate student, now a Ph. D. in Zoology, belonged to the acquaintance of the first author for a considerable long time, with the bonus of long discussion sessions in the Biology Bar, at the Federal University of Minas Gerais. Gilmar Ferreira do Carmo – the single M. Sc. student of M. L. Triques – and Dr. Mario Alberto Cozzuol, vertebrate paleontologist, also encouraged the authors in the beginning. Not all illustrations were used in the Portuguese or English editions, in their final versions. However, without the generous offerings of these colleagues, without their considerable altruism, we would never have started the strenuous process of writing and producing the Portuguese book. Without their help, the Portuguese edition would not have emerged and, consequently, neither this considerably expanded English edition.

Alfredo Hannemann Wieloch and Marcos Rodrigues also contributed greatly, with many ideas for the development of the book and its editing. Mauro Luís Triques also thanks them for the many hours of productive talks and, mainly, for their friendship and encouragement. Mário Cesar Cardoso de Pinna offered a clarification on the concept of "terminal" in phylogenetic systematics (any remaining errors are ours). Fabio Vieira helped with literature and many hours of productive conversation on alternatives in biological thinking. Amanda Ribeiro Cocovick helped edit the illustrations, although none of us were professionally trained in this field. Both Fábio and Amanda provided much encouragement, including support at crucial moments, and also many hours of pleasant and creative conversation.

Mauro Luís Triques considers this book to be the culmination of his career, synthesizing his collection of more important works and adding much to them. He would thus like to thank people who have been very important to him, from the beginning of his professional training, to those who helped him directly in concluding this book, and for maintaining motivation and a high spirit in this present moment of his life. His parents, José Walter Triques (who passed away on January 15, 2021) and Clelia Lygia Cerri Triques, have always supported him both in terms of professional education and mainly in relation to the formation of his character and provided invaluable emotional support.

His sister, Adriana Lúcia Cerri Triques, has been a constant presence and greatly encouraged the publication of this book. In critical emotional times through life, his family kept him standing. On the other hand, there is the funny side of Mr. and Mrs. Triques being forced to read repeatedly about dinosaurs to the boy before the age of literacy. Mauro is deeply grateful to his advisors during the times of under-graduation (Ricardo Macedo Corrêa e Castro), during the time of fulfilling his master's degree (Ricardo Sousa Rosa) and during his Ph.D. courses (Naércio Aquino Menezes), for the professional training received and friendship. The late Professor João Juarez Soares kindly received Mauro in his laboratory in the Botany Department of the Federal University of São Carlos, kindly permitting the use of a stereomicroscope for the under-graduation monograph on the fish skeleton. Ricardo Macedo first introduced phylogenetic systematics to Mauro, in an extra-class lecture, and permanently changed the professional direction of the 19-years-old student. José Carlos de Oliveira taught discipline in graduate project design, in the first conversation between both, of 15 minutes duration, in the Zoology Museum of the University of São Paulo, just before the beginning of M. L. Triques' M. Sc. thesis. In terms of vocational training, a special role was played by Professor Martin Lindsey Christoffersen, responsible for the enthusiastic and critical teaching of Hennig's phylogenetic systematics, to the then graduate student Mauro, who thanks Martin immensely. The exhilarating lectures given by Professor Christoffersen were never forgotten. Professor Martin Lindsey Christoffersen was the bulwark of the common work that led to the current book. Exaptation, adaptationism, criticisms on adaptationism, Blackburn's method, were never heard before, until Martin brought them under discussion. This also determined Mauro's career permanently, then with the age of 43. M. L. Triques also recalls the pleasant conversations with Martin in several snack bars and restaurants of Belo Horizonte, tasting the local cuisine and beer, while talking about life, Brazil and exaptations. The late Dr. Walter J. Bock very kindly sent his articles upon request, to a John Nobody from the end of the World; Mauro was moved to the point of accounting the event to his mother, who replied that it was a lesson to be learned. Bianca Machado Cerrutti had to listen to the desperate outburst of Mauro, considering the job on exaptation *versus* preadaptation, late at night, which is unforgettable. In the impossibility of individual thanks to all people, including the friendship, always fundamental, Mauro ends his words here.

Martin Lindsey Christoffersen would like to take this opportunity to thank his late father, Finn Elwarth Christoffersen, who departed from a full life at the age of 94 years, before the publication of our epic English edition. Finn very early in life taught Martin to 'value independent thinking', 'to believe in being different from others', and 'not to be intimidated by being different from the majority', even when such differences in opinions inevitably puts one into question, in contests, or conflicts with the dominant ideas. This synergy was rapidly shared by the authors, permitting the work to continue, despite the whirl of rejections and criticisms against the ideas developed herein. Of course, both families got their more than reasonable share of discussions and views from grandfather while working on fish and dinosaurs. My intellectual scientific father, Nelson Bernardi, was among the first pioneers returning to São Paulo, Brazil, after graduate courses abroad and contacts with the ideas of Willi Hennig. In his first course to five Ph.D. students in 1978, he introduced Phylogenetic Systematics to these students like a theologian converting his offspring to a new faith. With me, at least, he was fully successful. I almost threw out my ongoing Ph.D. thesis on geographical distribution of marine animals, because I gained the notion that this was not Historical Biogeography at all. Fortunately, I ended by concluding my descriptive project, gained my Ph.D., and waited to start my new career in Phylogenetic Systematics until after being appointed to my life job in Northeastern Brazil. The family at home, especially my wife Maria de Fátima Gomes de Lima Christoffersen, my grand-daughter Sofia Lindsey Christoffersen Melo Freire, then two and now eleven, and my sister-in-law, Eliana Gomes de Lima, followed never ending working spells on the book and related articles. Fátima was always patient with the long hours I spent on research. Sofia followed my explanations of what I was doing with interest, either on my lap or in front of the computer screen. Eliana, who after seeing me writing a revolutionary book along the years on the beach, during festivities, at most social gathering with the family, finally gave up asking when the book was going to be published, much before the Portuguese edition actually appeared. The scientific family at the University includes students in Palaeobiology classes and the research group in the lab (undergraduates, graduates, and postgraduates). This side of my relationships certainly sees me more than my family back home (although I do divide my time equally, from eight to eight at work, and from eight to eight at home). My professional family gets quite a lot of talk on evolutionary events in the present and the past, in class and at our regular four o'clock gathering for coffee in the lab. Contrary to the established gatekeepers of Science, responsible for accepting or rejecting publications, none of these folks at home and at work ever dissuaded me on the chosen research project.

We take the opportunity to thank the former couple Debbie McLennan and Daniel Brooks, esteemed affinity-induced friends of the senior author (Martin) during a week-long scientific event back in the 90's, for developing similar ideas along parallel lines. Their concept of co-option is similar to our concept of exaptation. Although pursued with different arguments, all four authors were mostly unaware of the published ideas of each other. The last book of Agosta and Brooks does make the third reference in the literature to our 2009 paper. This illustrates how Science often grows in parallel and develops from the insights of different research groups.

Specifically on the English edition, the authors would like to thank many encouraging people and publishing companies. Our contact with Brill editors started with Dr. J. Carel von Vaupel Klein. He edited a Chapter by the second author on the position of Pentastomida in a volume on Crustacea belonging to the series "Treatise on Zoology". He promptly and very kindly offered to forward our English version of the present work to the appropriate editors in Brill. This represented an enormous encouragement for the authors. Indeed, Dr. Von Vaupel Klein encouraged us with complimentary words concerning the original manuscript to Brill. He has provided excellent insights into the opening Chapters. Mr. Stefan Einarson was our subsequent contact at Brill and directed the conversation more specifically. He showed strong enthusiasm as the authors argued that this book is on biological evolution instead of being strictly a literary work on paleontology. This book has so far been seven years in construction at Brill, with a Covid pandemics slowing down the revision process considerably. Yet we have counted at all times with the technical help of the professional staff at Brill. We extend our thanks to Rosanna Woensdregt, in particular, for her support and patience. Although the book theme clearly includes both the subareas of zoology and paleontology, it probably reaches much further and deeper in displaying alternative ways for thinking in biology. Rosanna Woensdregt kindly and with special patience helped us with several aspects of editing the initial manuscript. Final steps for the construction of this book were provided by Associate Editor Simona Casadio and by Operations Manager, Savitha J. We are very grateful to the receptive and professional attitude of the staff we encountered at Brill.

On Illustrations

Public domain illustrations were fundamental for the project of producing this book. Wikimedia Commons contributed strongly by providing public domain images.

Biodiversity Heritage Library (https://www.biodiversitylibrary.org/) freely gives access to several publications, including public domain books (including text books), from which we could obtain public domain illustrations. Without such sources we would have had difficulties in producing our book. BioHeritageLibrary (BHL) makes evaluations on copyrights, indicating which literary works are in public domain. We were benefitted in obtaining illustrations free of copyright, such as from Goodrich (1909), Kingsley (1899), Newman (1923), and Woodward (1898), which are in public domain according to BHL. The BHL megaproject further permitted us to download these works.

Illustrations

Figures

1-1 A – Porbeagle shark, *Lamna nasus*, (Chondrichthyes). B – Tarpon, *Megalops atlanticus*, (Actinopterygii). C – Rock monitor, *Varanus albigularis*, (Lacertilia). D – Female newt, *Triturus cristatus*, (Lissamphibia). E – Alligator, *Alligator mississippiensis*, (Crocodylia). F – Pond-tortoise, *Emys orbicularis*, (Chelonia). G – Reconstruction of †*Archaeopteryx*, (Aves). †*Archaeopteryx* was chosen, instead of a modern bird, to show the original fingers in the wings, the absence of a beak and the presence of a backboned tail; these features are more easily compared with homologous characters noticeable in other vertebrates. H – Hyena, *Crocuta crocuta*, (Mammalia) 4

1-2 Cladogram for the large groups of Gnathostomata (vertebrates that shut their mouths by means of the mandible) 6

1-3 Phylogeny of large living groups of Actinopterygii 8

1-4 Phylogeny of large living groups of Sarcopterygii. Tetrapoda is enhanced in the phylogeny 10

1-5 Skeleton of European perch, *Perca fluviatilis* (Actinopterygii, Teleostei, Percomorpha) 13

1-6 Cranium of the European perch, *Perca fluviatilis* (Actinopterygii, Teleostei, Percomorpha) 14

1-7 Post-cranial skeleton of *Perca fluviatilis* 15

2-1 Illustration of non-rhipidistian sarcopterigians 29

3-1 Batfish *Ogcocephalus vespertilio* in dorsal (A) and frontal (B) view 37

4-1 Illustration of hydrodynamic vertebrates, of fusiform shape or very similar to fusiform shape 47

5-1 Cladogram of Osteichthyes with selected taxa, illustrating the origin of terrestriality in Tetrapoda (vertebrates with feet) 70

5-2 Reconstruction in lateral view of the "†osteolepiform" (Tristichopteridae, Sarcopterygii) †*Eusthenopteron foordi* 74

5-3 Reconstruction of the "†osteolepiform" †*Osteoloepis macrolepidotus* (Sarcopterygii) in lateral view 74

5-4 Lateral view of skeleton of the fossil coelacanth (Actinistia, Sarcopterygii) †*Undina* [=*Holophagus*] *gulo* 74

5-5 Ventral view of roof of the mouth, or palatum of the living Australian lung fish *Neoceratodus forsteri* (Dipnoi, Sarcopterygii) 75

5-6 Reconstruction of the roof of the mouth, or palatum, of the fossil lung fish †*Dipterus sp.* (Dipnoi, Sarcopterygii) 75

5-7 *Amia calva* (Halecomorphi, Actinopterygii), a North-American freshwater fish ecologically similar to the trahiras of South America 76
5-8 Reconstruction in lateral view of †*Hypsocormus insignis* (Actinopterygii), scales omitted 76
5-9 Reconstruction of the fossil lung fish †*Dipterus valenciennesii* (Dipnoi, Sarcopterygii) 77
5-10 Skeleton of the living African lungfish (Dipnoi, Sarcopterygii) *Protopterus annectens* 78
5-11 Ventral view of pelvic (or ventral) fins and pelvic girdle of the living Australian lung fish *Neoceratodus forsteri* (Dipnoi, Sarcopterygii) 82
5-12 Ventral view of pelvic (or ventral) fins and pelvic girdle of the "†osteolepiform" †*Eusthenopteron foordi* (Tristichopteridae, Sarcopterygii) 83
5-13 Inner view of skeleton of pectoral girdle of the right side of the "bichir", an African fresh-water fish (Cladistia, Actinopterygii) 84
5-14 Lateral view of the left fin skeleton of the fossil shark †*Pleuracanthus guaudryii* (Chondrichthyes) 85
5-15 *Microbrycon minutus*, holotype, 24 mm standard length (Actinopterygii, Teleostei, Ostariophysii, Characiformes) 87
5-16 Artistic reconstruction of †*Acanthostega* by Alessandra Amadori, based on Triques and Christoffersen (2017: 167, fig. 5-16) 87
5-17 Pelvic girdle (both sides illustrated) of fossil "amphibian" †*Mastodonsaurus giganteus* (Temnospondyli, Batrachomorpha, Tetrapoda) 88
5-18 Caudal vertebra of tuna fish *Thynnus vulgaris* in lateral view 89
5-19 Lateral view of a precaudal (abdominal) schematic vertebra of †*Hyloplesion* (Tetrapoda, Neotetrapoda, Batrachomorpha, †Microsauria) 90
5-20 Ventral view of roof of mouth, the palatum of the fossil amphibian †*Cyclotosaurus robustus* (Temnospondyli, Batrachomorpha, Tetrapoda) 95
5-21 Skeleton of chimera (*Chimera monstrosa*, Chondrichthyes) 101
5-22 Cladogram of *fig.* 5-1, inverted, that is, rooted with †Baphetidae + the crown group 109
5-23 Simplified phylogeny based on *fig.* 5-22, given to facilitate general visualization 110
5-24 Phylogeny derived from *fig.* 5-23, which differs for permitting the inclusion of any possibilities of phylogenetic relationships among any subgroups of sarcopterygians except †Baphetidae and the terrestrial vertebrates (crown group) 111
5-25 Inverted cladogram alternative to *figs.* 5-22, 5-23 and 5-24 112

Tables

2-1 Geological scale of the Earth 22

Notes for Reading This book

Professionals not accustomed to reading specialized technical texts will benefit from some clarification. References to the scientific literature is in a formal way, thus permitting that the reader will be directed to the precise source of our citation. In an imaginary example, instead of saying that the first and second author of this book published a scientific article in 2009, saying that on page 37 "such a group is monophyletic", as can be seen in their figure number 2, all the above information is presented in a concise way: "such a group is monophyletic" (Triques and Christoffersen 2009: 37, *fig.* 2). With such customary convention we try to avoid distorting the words of the authors on which we have relied upon: we try to quote them as literally as possible. By doing so, we make the most accurate statement as possible, so that any reader can trace our sources in detail to the correct page, when applicable, as explained above.

Latin expressions help us to present, in abbreviated form, what we wish to indicate, being based on Papavero (1994: 148–150). Citations with more than two authors are made using the Latin expression *et alii*, abbreviated as *et al.* (hypothetical example, Triques *et al.* 2001). In this example, we do not wish to indicate any specific page, or figure, we are only referring to a published paper or book. "Paper" means a scientific article. Other words and their abbreviations used in this book are as follows: *apud* (*ap.*), in the publiction of; *citatus* (*cit.*), cited; *exempli gratia* (*e.g.*), for example; *figure, figures* (*fig.*), figure; *id est* (*i.e.*), that is; *ibidem* (*ib., ibid.*), the same, in the same place [used to indicate repetition in the citation of a publication, in a bibliography or in the text]; *opere citato* (*op. cit.*), in the cited paper; *pagina* or *paginae* (*p., pp*), page or pages; *sensu*, in the sense of, according to. These words or their abbreviations are written in italics, to indicate that they belong to another language. Mentions for "millions of years before present " are indicated as "Myr", for millions of years ago.

This book deals with taxa that no longer have living members, that is, that are no longer extant, and are called fossil taxa. Other taxa present both fossil and living (extant) forms; as a whole, this type of taxon (a group of organisms, considering it as an evolutionary unit – in plural form, taxa) is referred to as living, because it contains living representatives. For the reader to know the situation to which a taxon belongs along the text, the symbol "†" is inserted before every fossil taxon. This symbol may also be inserted before or after fossil taxa appearing in the branched diagrams we use in the book, which are of a specific kind, called cladograms. A glossary is present before the references.

CHAPTER 1

Phylogenetic Systematics or Cladistics, Introductory Information

After the success of the ideas of Darwin, researchers that work with the classification of animal species (zoological systematics) started rethinking ways of seeing characters, now within the evolutionary paradigm.

The core of our argument, in favor of exaptations in the evolutionary passage of vertebrates from water to land, is based on the basic concepts of phylogenetic systematics, also known as cladistics. These concepts need to be mastered by the reader. In the glossary we present the relevant terminology.

1 From Darwin to Hennig

The concepts of phylogenetic systematics derive from the theory of the origin of species by natural selection, by Charles Darwin. Willi Hennig (1950, 1966) advances the seminal work of Darwin, establishing the method of Phylogenetic Systematics, later formalized as Cladistics. Systematization would be a more adequate term than the classical term "classification". Phylogenetic systematics is based on System Theory. This view uses individuals, which are entities delimited in space and time, such as organisms, that are born, live, and die in a certain area and exist during a certain period of time. Besides these paradigmatic individuals (the organisms), monophyletic groups also represent historical entities that behave like individuals. The German entomologist Willi Hennig proposed, already in 1950, that living systems, rather than abstract and non-natural classes, are individuals characterized by shared exclusive ancestry, inferred from evolutionary novelties. Since the advent of phylogenetic systematics by Hennig, not only species, but also supra-specific taxa have the status of evolutionary units. For example, while Metazoa originated in the ocean in the Pre-Cambrian era, its subgroup Tetrapoda only conquered land much later, in the Paleozoic. Mammalia appeared only in the Mesozoic, radiating spectacularly in the Cenozoic, after the extinction of dinosaurs.

Pre-evolutionary classification used classes (universal entities not delimited spatio-temporally, recognized according to their definition). For example, the chemical elements are classes defined by their atomic number. Gold represents the same substance, whether found on Earth, or in any other location

 | DOI:10.1163/9789004684911_002

in the universe. Books in a library, as another example, are classes and can only be classified. Several classifications of books are possible: by author, by subject, by editor, by size, by language, etc. Utility of these multiple possible classifications may be the only criteria for choosing among alternative classifications. Pre-evolutionary biological classifications tended to be anthropocentric, and only useful animals or animals of interest to man tended to be studied and classified into detailed artificial groups.

Phylogenetic systematics avoids the essentialist and typological connotations of classes of organisms (as we will see below), which in pre-evolutionary times were defined by shared similarities and differences. The classes of the Linnean system of classification are inadequate for a transformist (evolutionary) outlook of the world. Hennig argued that there is only one valid natural system of living beings. The search for this unique natural system now guides the activity of systematists. Because diversity is a product of organic evolution, the systematization of living beings must necessarily be phylogenetic (cladistic).

The evolutionary ideas of Darwin are of fundamental importance for biological Science. The comprehension of the effects of evolutionary thinking was not and is not automatic, much less simple and direct. Evolutionary principles and methods were deduced very slowly (Christoffersen 2011). Much research and debate are still necessary, in order to induce evolutionary thinking into each area of biology. Phylogenetic systematics represents a revolution within biological systematics, emerging from the proposal by Darwin of the origin of species by natural selection. Phylogenetic systematics appeared only a century after the publications of Darwin. It substitutes the millennial Aristotelian paradigm of the existence of immutable entities defined by essences and substitutes this world-view with a transformist paradigm. A systemic ordination is more adequate for the evolutionary vision introduced by Darwin. Classifications by similarities and differences (typological classifications) characterize all cultures since civilization. Because of this situation, classifications by overall resemblance became practically intuitive. Thus, the effort to systematize diversity into hierarchically organized sets may become quite difficult and anti-intuitive for many students and scientists, still immersed in the dominant classification paradigm (developed before the ideas of Darwin). For example, the traditional classification divides animals into invertebrates (animals without vertebrae) and vertebrates (animals with vertebrae), and then subdivides the vertebrates into fish (vertebrates living in water), Amphibia (vertebrates that still depend on water to complete their reproductive cycle), reptiles (vertebrates that conquered land, but still have the body covered with scales), birds (endothermic vertebrates with feathers, that conquered the aerial niche), and mammals (endothermic vertebrates with hairs, that produce milk to feed their brood). This is binary reasoning, making a distinction between organisms that possess certain characters from other organisms that do not possess these traits. Which characters (or character states) are ancient in evolution (ancestral condition) and which are new (derived condition)? Vertebrae represent the evolutionary novelty in the evolution of animals (derived condition), while the absence of these structures is the ancestral condition. To retrieve the evolutionary knowledge, and to discover novelties (derived conditions), is the path to recover the evolutionary path of organisms. The ancestral condition does not indicate evolution, while the derived condition shows exactly which evolutionary path was taken. Vertebrata is a group defined by derived characters (vertebrae), apomorphies. Consequently, it represents a natural group in evolution. On the other hand, invertebrates are defined by the absence of a derived character (vertebrae), a plesiomorphy, and thus do not form a natural group from an evolutionary point of view: it represents an artificial group. Finding derived characters is thus the same as finding natural groups from an evolutionary perspective. Derived characters are the sole key to natural groups. The ancestral condition is uninformative regarding evolutionary relationship, and produces artificial groups based on perceived similarities, at least to the human eye. Regardless of any perceived similarities among invertebrates, compared to vertebrates, this grouping (Invertebrata) was established with no regards to evolution, despite the fact that the absence of vertebrae is a similarity of invertebrates. In the words of Christoffersen (2011), invertebrates should not be recognized by the absence of vertebrae, because this similarity has no correspondence to the evolutionary process (similarity is evolutionarily meaningless). All ancestral similarities are meaningless for the recovery of evolutionary pattern. On the other hand, we may infer that the group of vertebrates has a common exclusive ancestor, because of the unique sharing of vertebrae – that do not exist in any other living or fossil organism (Christoffersen 2011).

In phylogenetic systematization, the perception of taxa grouped by similarities is substituted by the perception of taxa as groups that share characters that gradually appear during the evolutionary process. Thus, marine invertebrates such as ascidians are classified together with their vertebrate relatives, in the more inclusive taxon Chordata. Some fish such as coelacanths are classified together with their tetrapod relatives, in the more inclusive taxon Sarcopterygii. Some Amphibia, such as the Lissamphibia, are classified together with their amniote relatives in Tetrapoda (see *fig.* 1-1 for some representatives). Crocodiles and their kin are classified together with their bird relatives, in the taxon Archosauria. And some fossil Therapsida, traditionally classified as fossil reptiles for not yet possessing most mammalian characters, are considered

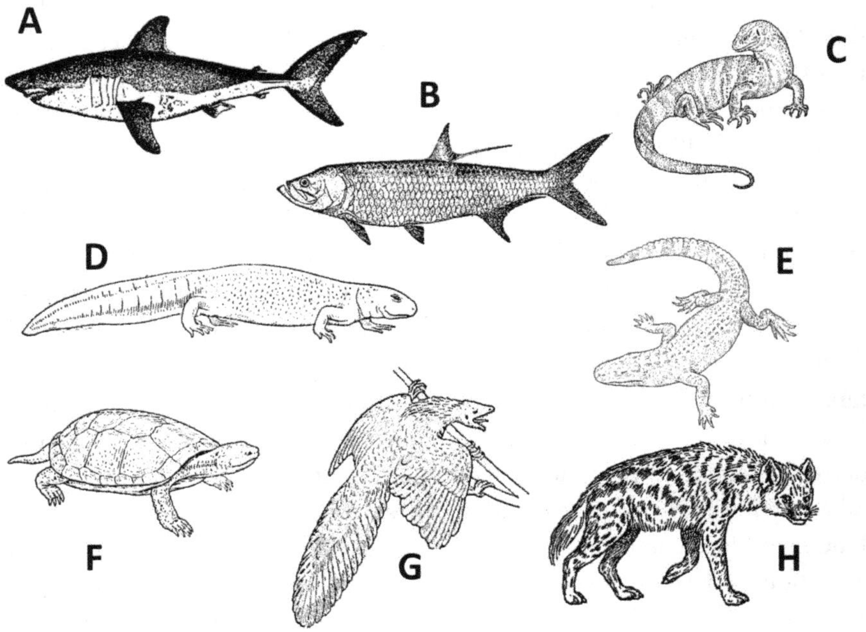

FIGURE 1-1 A – Porbeagle shark, *Lamna nasus*, (Chondrichthyes). B – Tarpon, *Megalops atlanticus*, (Actinopterygii). C – Rock monitor, *Varanus albigularis*, (Lacertilia). D – Female newt, *Triturus cristatus*, (Lissamphibia). E –Alligator, *Alligator mississippiensis*, (Crocodylia). F – Pond-tortoise, *Emys orbicularis*, (Chelonia). G – Reconstruction of †*Archaeopteryx*, (Aves). †*Archaeopteryx* was chosen, instead of a modern bird, to show the original fingers in the wings, the absence of a beak and the presence of a backboned tail; these features are more easily compared with homologous characters noticeable in other vertebrates. H – Hyena, *Crocuta crocuta*, (Mammalia).

DRAWINGS FROM NEWMAN (1923), EXCEPT FOR THE PORBEAGLE SHARK, WHICH IS BASED ON GOODRICH (1909). ALL IMAGES ARE IN THE PUBLIC DOMAIN. SEE TEXT, FOR FURTHER INFORMATION ON TAXA AND CHARACTERS.

reptiliform mammals instead of mammalifom reptiles. The phylogenetic system substitutes the linear division of the diversity of living beings, characterized by the presence or absence of similarities, by a hierarchical system of organisms, characterized by evolutionary novelties (innovations) that result from the evolutionary process.

2 Which Animals Are We Dealing With?

Even the graduate biologist may find it difficult to identify the taxonomical level at which the transition from water to land occurred in the vertebrates.

Some vertebrates do not have mandibles, so that they do not close their mouths with an upward movement of the chin. Lampreys (Petromyzontiformes or Petromyzontoidea) have a circular mouth that adheres to the victim, rips out the victim's skin by means of the toothed tong, and then sucks the blood, skin and muscle fibers, during feeding. Hagfishes (Myxiniformes or Myxinoidea), strangely for us, close and open their mouths laterally, not in up-and-down movements like ours. Indeed, Myxiniformes have a skull, but not vertebrae. In this book we go directly to the vertebrates having a mandible, that is, those able to open and shut their mouths with an up-and-down movement of the lower jaw or mandible, when they want to, or need to. We humans do this dislodging the chin, so that the lower lip finds the upper lip, and our lower teeth contact the upper teeth, enabling us to chew our food.

In mammals, the mandible is formed by a single bone on each side, that articulates with the skull near the ear. You may feel the articulation functioning if you place your indicator finger in front of the ear, pressing the skin, while opening and shutting the mouth. In humans, this bone fuses with its pair from the other side of the head, forming the chin. Our mandible thus contains a single bone. Animals with mandibles are grouped into the Gnathostomata. "Gnathos" means mandible (the set of bones that form the inferior, movable part of the mouth) and "stoma" signifies mouth, in Greek.

In *fig.* 1-2 we have a cladogram with the large groups of living and fossil gnathostomes. †Placodermi is the sister group of the remaining groups, that are named Eugnathostomata. These include the Chondrichthyes (sharks, dog-fish, saw-fish, saw dog-fish, rays [including mantas] and chimeras) and their sister group, the Teleostomi. This latter taxon contains †Acanthodii and its sister group, the Osteichthyes. The osteichthyans are popularly named "bony fishes", in order to discern them from the "cartilaginous fishes" (Chondrichthyes or chondrichthyans), but this distinction is "functional" only in terms of living gnatostomes. †Placodermi and †Acanthodii are heavily ossified, at least in the more superficial (dermic or dermal) area of their skeletons. Osteichthyes are then divided into Actinopterygii and their sister group, the Sarcopterygii. It is within the Sarcopterygii that the transition of life in water to life on land occurred. The event happened only once within the group (*e.g.* Clack and Ahlberg 2016, Clack *et al.* 2016). This means that the totality of terrestrial living and fossil vertebrates (those that can walk, breath and feed on land) form a monophyletic group, having a single common exclusive ancestor of their own. The cladograms presented by Clack *et al.* (2016) permit the interpretation that terrestriality emerged twice, but only in the most derived portion of the Sarcopterygii lineage. Indeed, this event is indicated as only occurring in the most derived portion of the tetrapod lineage. This possibility was already indicated and discussed by Triques and Christoffersen (2015). However, the possibility

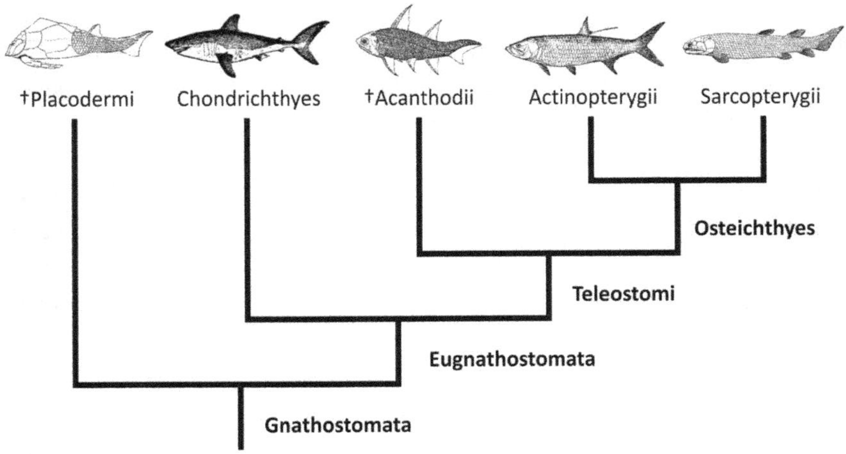

FIGURE 1-2 Cladogram frequently accepted (*e.g.* Pough *et al.* 2005) for the large groups of Gnathostomata (vertebrates that shut their mouths by means of the mandible). The †Acanthodii is considered paraphyletic and its position is considered *incertae sedis* in Pough and Janis (2019: 96, *fig.* 6-1).
FIGURES ARE FROM GOODRICH (1909) EXCEPT FOR THE ACTINOPTERYGIAN, WHICH WAS TAKEN FROM NEWMAN (1923). ALL IMAGES ARE IN THE PUBLIC DOMAIN.

that terrestriality appeared twice in the evolution of the Tetrapoda does not change our proposals in this book. We continue to support the alternative position that terrestriality in Tetrapoda represents a unique, unparalleled evolutionary event. We will return to this point in Chapter 5.

Although a very interesting group in itself, Chondrichthyes are phylogenetically distant from Sarcopterygii and do not have representatives with any link to terrestrial environments, such as leaving the water to capture prey and then returning to their aquatic realm. We will thus not consider this group further.

†Placodermi were a diverse group from a morpho-functional point of view, but is also cladistically distant from Sarcopterygii, so they will not be presented in detail. They had a heavy cephalic carapace built from bone tissue, articulated to a carapace of similar nature covering the thorax. It is believed that such heavy animals inhabited the water bottom or lived just above the bottom. In *fig.* 1-2, †Placodermi appear as a monophyletic group, as in many textbooks. However, several recent papers suggest it may be a paraphyletic group, that would then include the Eugnathostomata. This may have influence on one of our arguments against adaptationism and will be discussed in detail at the appropriate moment.

†Acanthodii were little diversified morpho-functionally, containing animals with spines in all their fins, excepting the caudal one. They were much lighter than the †Placodermi, more hydrodynamic and not likely benthonic,

most probably swimming above the bottom. Some of these animals were filter-feeders. In *fig.* 1-2, †Acanthodii appear as a monophyletic group, as in many textbooks. Following the case of †Placodermi, some recent papers suggest they may be a paraphyletic group that includes Chondrichthyes, and maybe even Osteichthyes. Again, this may influence one of our arguments against adaptationism and will be discussed in detail later.

In Actinopterygii, the diversity is very large, even taking into account only the living representatives. Estimates indicate that the actinopterygians contain approximately 30,000 species, roughtly half the number of living species of vertebrates of the world (Friedman and Giles 2016). Pough and Janis (2019: 2, *fig.* 1-1) show Actinopterygii as having 32,082 species. A cladogram of the main living groups of actinopterygians is seen in *fig.* 1-3. Most representatives belong to the Teleostei, indicated in the figure.

Sturgeons (Chondrostei), are known for their ova, the famous caviar, an expensive food considered a delicacy. In Teleostei, Osteoglossomorpha include the largest freshwater fish from South America. The figure illustrates an electric freshwater fish from Africa. Elopomorpha include morays, eels, tarpons, and others. In some recent phylogenies, Elopomorpha appear as the sister group of the remaining Teleostei (*e.g.* Alfaro *et al.* 2009). Clupeomorpha are the group to which sardines and herrings belong. Salmoniformes include salmons and trouts. Pough and Janis' scheme (2019: 134, *fig.* 9-1) differs from our *fig.* 1-3 in that Lepisosteiformes and Amiiformes are sister groups, Clupeomorpha is the sister group of Ostariophysii, and Osteoglossomorpha and Elopomorpha are *incertae sedis* groups.

Ostariophysi include 27% of the teleosts of the world, and about 65% of the world's fresh-water species (Nelson, 1994). According to Nelson *et al.* (2016: 175), 10,388 fish species were recognized as belonging to Ostariophysi, representing 28% of the known fish species in the world, while accounting for about 68% of the freshwater species, percentages only slightly different from the evaluation by Nelson (1994). Within Ostariophysi, we encounter the carps and their relatives (Cypriniformes). Siluriformes include catfish, mailed catfishes, and others. Gymnotiformes are the fresh-water electrical fish of South America and are devoid of direct relationships with the African electric species (that belong to the Osteoglossomorpha or represent African Siluriformes). They include the electric eel from South America and the less known knifefishes. Characiformes contain several South American endemic groups: tetras, neons, piranhas and many others. There are also African representatives such as the river tiger-fish.

In some more recent phylogenies, Clupeomorpha appear as the sister group of the Ostariophysii (*e.g.* Alfaro *et al.*2009), but see Nelson *et al.* (2016: 161) for an alternative proposal.

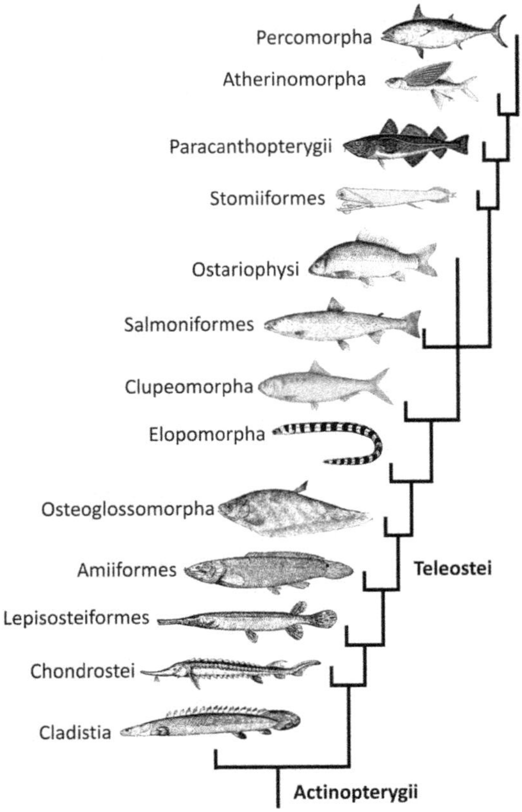

FIGURE 1-3
Phylogeny of large living groups of Actinopterygii.
SIMPLIFIED FROM LAUDER & LIEM (1983). SEE POUGH AND JANIS (2019) FOR A DIFFERENT TOPOLOGY. DRAWINGS ARE AS FOLLOWS. *POLYPTERUS BICHIR* (CLADISTIA), FROM GOODRICH (1909). *ACIPENSER RUTHENUS* (CHONDROSTEI), FROM NEWMAN (1923). *LEPISOSTEUS PLATYSTOMUS* (LEPISOSTEIFORMES), FROM NEWMAN (1923). *AMIA CALVA* (AMIIFORMES), FROM NEWMAN (1923). *NOTOPTERUS SP.* (OSTEOGLOSSOMORPHA), FROM GOODRICH (1909). *GYMNOTHORAX WAIALUAE* (= G. *RUEPPELLIAE*) (ELOPOMORPHA), FROM NEWMAN (1923). *CLUPEA ALOSA* (= *ALOSA ALOSA*) (CLUPEOMORPHA), MODIFIED FROM GOODRICH (1909). *SALMO SALMO*, (= *S. SALAR*) (SALMONIFORMES), FROM GOODRICH (1909). *CYPRINUS CARPIO*, (OSTARIOPHYSII), FROM GOODRICH (1909). *MALACOSTEUS INDICUS* (= *M. NIGER*) (STOMIIFORMES), MODIFIED FROM GOODRICH (1909). *GADUS MORHUA* (PARACANTHOPTERYGII), MODIFIED FROM KINGSLEY (1899). *EXONAUTES GILBERTI* (= *HIRUNDICHTHYS RONDELETII*) (ATHERINOMORPHA), FROM NEWMAN (1923). *THUNNUS* SP. (PERCOMORPHA), FROM JORDAN & EVERMANN (1902). ALL IMAGES ARE IN THE PUBLIC DOMAIN.

Stomiiformes, Aulopiformes, and Myctophiiformes are deep-water marine teleosts, many of them with specific patterns of photophores to provide an identifiable (among individuals of the same species) means to find a sexual partner; also, many species have large eyes and mouths. Paracanthopterygii include the cod fish, hakes, angler-fish, bat-fish, and others. Atherinomorpha include flying fish, guppies, and others. Percomorpha has about 15,500 species (*e. g.* Alfaro *et al.* 2009) and may not be well circumscribed evolutionarily. They may include Atherinomorpha and Mugilomorpha (the mullets). Most coral reef fish are usually placed in this taxon, such as the clown fish, puffer, grouper, flatfish, the famous periophthalm of mangroves or mudskipper (*Periophthalmus*), as well as tuna fish, sword-fish, and the strange moon-fish (*fig.* 1-3), an enormous diversity which we are unable to deal with here. We will likewise not detain ourselves much with Actinopterygii, which will be treated properly in Chapter 4, but not in terms of each of all of its modern groups. The focus of this book is on the evolutionary transition from water to land, and such a transition has not occurred in Actinopterygii, but only in their sister group, the Sarcopterygii.

In *fig.* 1-4 we present the phylogeny of the large groups of living Sarcopterygii. Fossil groups were omitted. The coelacanths (Coelacanthiformes or Actinistia) are frequently accepted as being the sister group of the remaining taxa. Considering only living forms, the lungfish (Dipnoi) form the sister group of the Tetrapoda ("tetra" means four and "podo" stands for foot, in Greek), which represent the vertebrates having feet (our hands represent feet that became adapted for clinging). It is precisely at this point of the cladogram that the evolutionary passage to land occurred. Living Tetrapoda do not have to return to water in order to live (although some still have to do so in order to breed). As we shall see in Chapter 5, the situation becomes much more complex and interesting when the fossils are included in the cladogram. Several evolutionary conclusions will be possible based on these data (see Chapter 5).

Returning to the living forms of Tetrapoda, we have the living amphibians (Lissamphibia) as the sister group of the remaining taxa placed in Amniota. This name relates to the presence of an amnion, which is a membrane that covers the developing embryo. Amniota include vertebrates that may reach great size as adults and which are able to colonize drier environments than the Lissamphibia, although this is not entirely uncontroversial. Mammalia include the mammals. Chelonia (some authors prefer Testudines) include turtles and tortoises. Crocodilia include alligators, the gavial, crocodyles and others. It represents the sister group of Aves among living forms. Together, they both form the sister group of Lepidosauria, which include lizards and one of their subgroups, the serpents. In Pough and Janis (2019: 163, *fig.* 10-2), Testudines or Chelonia appear as the sister group of archosaurs, that include Crocodilia and Aves. We found contradictory spellings for taxonomic groups above the generic level *Crocodylus*: Crocodylia, Crocodylidae, Crocodyliformes, crocodyloid, Crocodyloidea, Crocodylomorpha (*e.g.* Pough and Janis 2019). We also found Crocodilia, and crocodilian (*e.g.* Romer 1966). As we do not aim to solve nomenclature issues in this book, we feel free to use both spellings in it, instead of choosing one.

Phylogenetic systematics has recovered the genealogical relationships of vertebrates in remarkable detail. As a consequence, it is very difficult to accommodate the Linnean categories in this system. The seven initial categories originally introduced by the great Swedish taxonomist Carl Linnaeus were subsequently subdivided (for example, superorder, order, suborder,

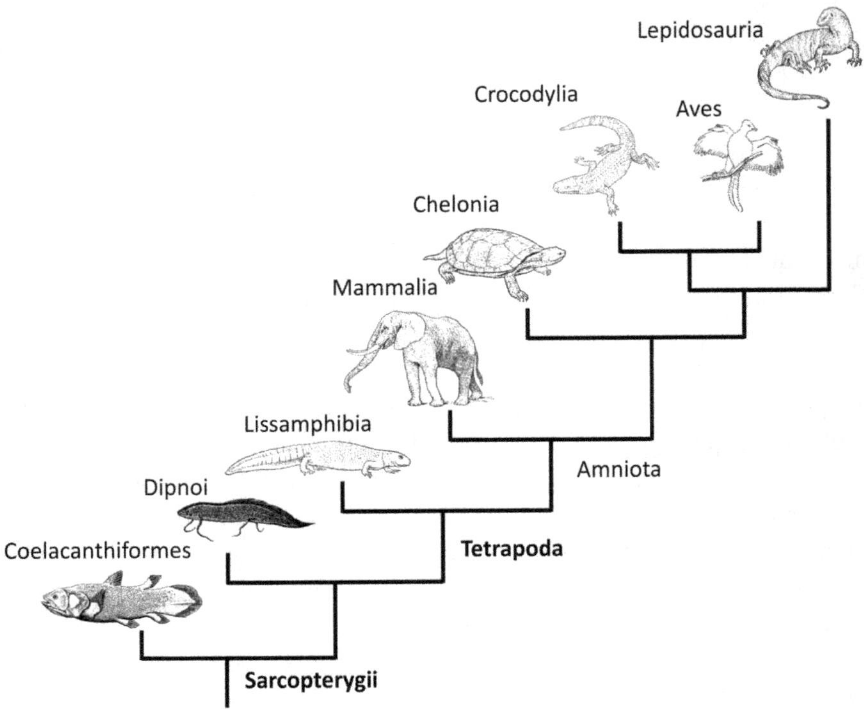

FIGURE 1-4 Phylogeny frequently accepted of the large living groups of Sarcopterygii (*e.g.*, Clack, 2002, 2012; Pough *et al.*, 2005). Tetrapoda is enhanced in the phylogeny. See Pough and Janis (2019: 162, *fig.* 10-1) for Chelonia as *incertae sedis.* ILLUSTRATION CREDITS ARE AS FOLLOWS. *LATIMERIA CHALUMNAE* (COELACANTHIFORMES), FROM HTTPS://COMMONS.WIKIMEDIA.ORG/WIKI/CATEGORY:LATIMERIA_CHALUMNAE#/MEDIA/FILE:COELACANTH-BGIU.PNG ATTRIBUTED TO BOGDAN; SEE ALSO HTTPS://COMMONS.WIKIMEDIA.ORG/WIKI/FILE:COELACANTH.PNG BY ROBBIE CADA (MAY 21TH, 2018). *PROTOPTERUS ANNECTANS* (DIPNOI) FROM KINGSLEY (1889). *TRITURUS CRISTATUS* (LISSAMPHIBIA)FROM NEWMAN (1923). *EMYS ORBICULARIS* (CHELONIA) FROM NEWMAN (1923). *ELEPHANS AFRICANUS* (= *LOXODONTA AFRICANA*) (MAMMALIA) FROM NEWMAN (1923). *ALLIGATOR MISSISIPPIENSIS* (*CROCODYLIA*) FROM NEWMAN (1923). †*ARCHAEOPTERYX* (AVES) FROM KINGSLEY, J. S. (1899). *VARANUS ALBIGULARIS* (LEPIDOSAURIA) FROM NEWMAN (1923). ALL IMAGES ARE IN THE PUBLIC DOMAIN.

infraorder, etc.) even before cladistic analyses became routine, but even at the time quickly became insufficient. Thus, we simply do not use such categories in this book. For example, it is not important whether Mammalia are considered a class, superclass, subclass, infraclass, division, or phylum: any one of these categories is subject to criticisms due to their subjectivity.

We may also note that, as Tetrapoda are the sister group of a taxon with the shape of a "fish", both taxa forming the sister group of another taxon with the shape of a "fish", and thus successively down to †Placodermi, which are then the sister group of the remaining Gnathostomata, the taxon Pisces, as conventionally used, is a paraphyletic group, because it arbitrarily excludes Tetrapoda. Thus, the term Pisces is not valid under the cladistic paradigm. It would only become acceptable if it were redefined to include all of the Tetrapoda, including human beings, within Pisces.

3 The Skeleton

Modern technologies have suggested that it is possible to study the soft parts of fossil vertebrates (*e.g.* Maldanis *et al.* 2016). Yet basically the vast majority of known data for the origin of terrestriality is based on information from the skeleton. For the sake of keeping the focus of the book on its basic theme, we will address here only the skeletal characters that we actually use in the book: ribs, fins (locomotor appendages, paired and median in the body), characters linked to the functional neck (series of opercular bones, upper part of the pectoral girdle and notochord) and aspects of the vertebral column. We will base our presentation on the skeleton mostly on Bond (1996) and Weitzman (1962) for teleosts, in a simplified way.

The skeleton provides mechanical support for the soft parts of vertebrates, and also functions as a support system for the muscles, that move the animal and its parts independently of each other. Much of the body form is due to the skeleton, including the outer shape, as may be observed in *fig.* 1-5, of a teleost.

In *fig.* 1-5 we perceive that the anterior region (that is positioned in front, when the animal moves forward) of a vertebrate with the form of a "fish" (in this case, a teleost) is formed by the cranium. The cranium (*figs.* 1-5 and 1-6) encounters most resistance to water when the animal moves. Thus, it has hydrodynamic importance. At the same time, the cranium protects the cerebrum and many sensory organs positioned on the head, both against the water pressure when the body is moving, and against collisions or even attacks from other animals. In *fig.* 1-6 the orbit, that rounded or circular opening in the cranium, is noteworthy, and this is where the eye is placed. It is also through the cranium, by way of the mouth, that water enters the body, for breathing and feeding. To open the mouth, the mandible is lowered. The mandible or lower jaw is formed mainly by the dentary bone (D in *fig.* 1-5 and D in *fig.* 1-6), which is the bone that supports the lower teeth. The water used for respiration exits the body by the opercular opening, which requires a lateral displacement of the lateral wall of the head. The bones responsible for this movable lateral wall (called the operculum) are the opercle, or opercular bone (O), subopercle (SO), interopercle (IO), located anteriorly to the SO, and the preopercle (PO), placed in front of the opercle. This set of four bones (opercular apparatus) also protects the branchial (inside of the skull) apparatus laterally. For the emergence of the functional neck, some of these bones were lost, as we will see in Chapter 5. The bone element that gives support to the opercular apparatus is the hyomandibular (HM), which is an elongate bone extending down and forwards, touching the quadrate bone (Q) by means of an intermediary bone, the symplectic (SP in *fig.* 1-6; SM in *fig.* 1-5). The hyomandibular connects part of the cranium dorsally – that portion which covers the cerebrum and at the same time protects the inner ear region and named neurocranium – down to the quadrate, ventrally. The quadrate (Q), then, anteriorly has a convex rounded facet that contacts the concave rounded facet of the angular or articular, forming a hinge (ball-and-socket articulation). This arrangement finally permits the up-and-down movement of the lower jaw in order to close and open the mouth. At the same time, the hyomandibular supports the branchial apparatus internally (not shown). The lower jaw articulates with the quadrate, through its anterior socket-shaped facet described above, as can be seen in *fig.* 1-6. The hyomandibular is thus responsible for suspending the lower jaw. The connection of the pectoral girdle with the cranium is located above the operculum. It includes the posttemporal (PTO in *fig.* 1-5 and PTP in *fig.* 1-6) and extrascapular bones in front (not illustrated in this figure). The remaining bones illustrated in *fig.* 1-6 will not be discussed further in this book.

The post-cranial region includes the remaining part of the skeleton (*fig.* 1-7). The pectoral girdle is formed by several skeletal elements, being indicated in *fig.* 1-7 as PG (pectoral or scapular girdle). Ventrally, there is the coracoid (CO), and above it, are placed the vertically elongate cleitrum (CL), supra-cleitrum (SCL) and, finally, the scapular bone or scapula (not indicated). Behind these bones are one or several post-cleitral bones (PCL). Above them all, but still a part of the pectoral girdle, is a posttemporal (illustrated together with the cranium). Beyond this (fixed to the cranium) is the extrascapular (not illustrated). The pectoral fin rays (P) are connected to the girdle by skeletal elements called radials (RAD). Other skeletal radial elements support the spines (ESP) on the dorsal (D) and anal (A) fins, and also the rays (RA) of these fins. Fin rays are segmented, elongated elements, which support the fin membrane. The spines are rays in which segments have fused, resulting in hard, pointed elements. One pelvic bone (OP in *fig.* 1-7) sustains the pelvic or ventral fin (V). The ventral (V) and caudal (C) fins also have rays. The pelvic bone may be represented by other, homologous elements, in sarcopterygians.

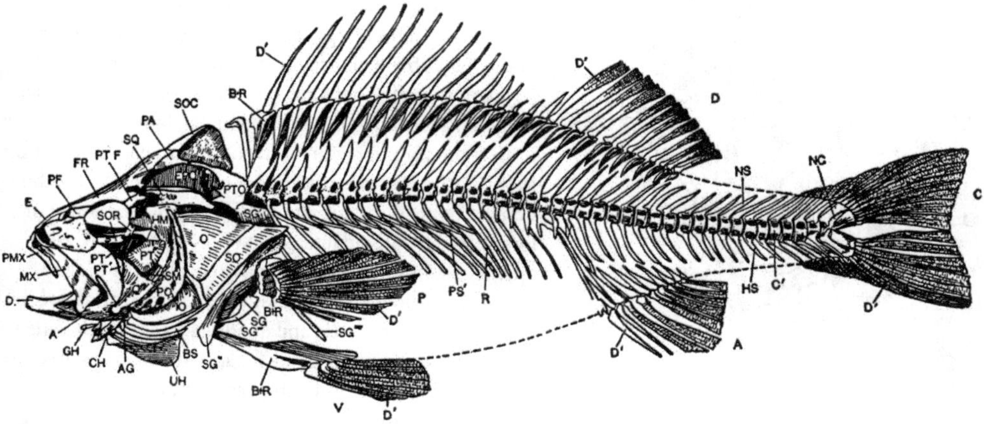

FIGURE 1-5 Skeleton of European perch, *Perca fluviatilis* (Actinopterygii, Teleostei, Percomorpha), in lateral view Anterior region on left side. A, anal fin. B+R, skeleton for spine and fin rays. BS, branchiostegal rays. C, caudal fin. C′, vertebral center, which is the cylindrical portion of the vertebra. D, dorsal fin, in this case with anterior part formed by spines (hard, non-segmented structures), and the posterior part formed mainly by rays (flexible, segmented structures). D., dentary bone, bearing the lower jaw teeth (not illustrated). FR, frontal. GH, glossohyal, the tongue bone. HS, haemal spine, below vertebral center. IO, interopercle. MX, maxillary. NS, neural spine, above vertebral center. O, opercle, (composing the "operculum" with subopercle, interopercle and preopercle) that displaces laterally to open the opercular aperture, permitting the passage outward of water used for respiration. P, pectoral fin, anchored in proximal radials, indicated in this figure as B+R. PMX, premaxillary (teeth not illustrated). PO, preopercular. PA, parietal. PTO, post-temporal (of pectoral girdle), in front of which lies (but not illustrated) the extra-scapular bone, which connects the cranium to the pectoral girdle. SO, subopercle. V, ventral or pelvic fin, in this case connected to the pelvic girdle, indicated in this figure as B+R.

FROM DEAN (1895). SEE A SLIGHTLY DIFFERENT AND MORE COMPLETE SKULL NOMENCLATURE IN *FIG*. 1-10. IMAGE IN THE PUBLIC DOMAIN

The vertebral column is formed by individual vertebrae. These have a vertebral centrum (C'/VC respectively in *figs*. 1-5 and 1-7), a neural spine (NS), which protects the dorsal nerve chord dorsally, and below a hemal spine (HS), that protects the large veins in the caudal region. Ribs (R) directed ventrally protect the viscera and are the pleural ribs. There may also be small ribs directed dorsally (PS'/PR), also called epipleurals. The vertebral centrum is a short bony cylinder. From it emerges, on each side, and upwards, a thin laminar bone. These converge toward the median plane and form the neural arch. The set of neural arches form the neural canal, which protects the dorsal nerve chord. The neural spine (NS) arises from the confluence of the two laminar walls. Ventrally a similar formation is present, originating the hemal arch, which protects the blood vessels. The hemal spines (HS) originate from these arcs.

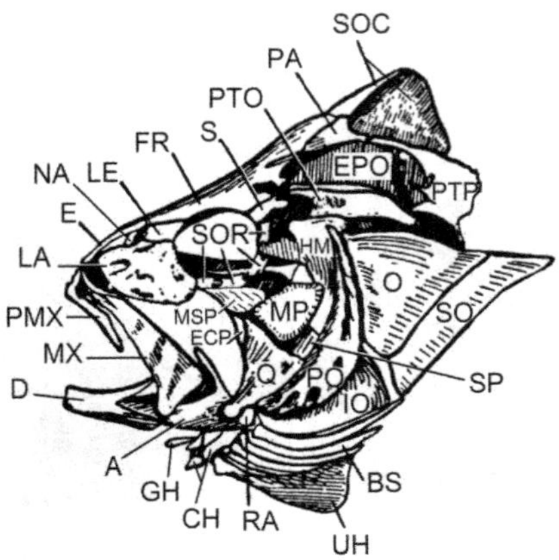

FIGURE 1-6

Cranium of the European perch, *Perca fluviatilis* (Actinopterygii, Teleostei, Percomorpha). Teeth not illustrated. Anterior region to the left. A, articular, a bone of the mandible that articulates with the quadrate bone (Q) permitting the opening and closing of the mouth. BS, branchiostegal rays; these are elongated laminar bones that protect the ventral region of the head. CH, cerato-hyal, that supports the ventro-lateral bones of the head (BS). D, the dentary bone, that supports the teeth of the mandible or lower jaw. E, ethmoidal region. ECP, ectopterygoid. EPO, epiotic. FR, frontal bone. GH, glosso-hyal, is the tongue bone. HM, hyomandibular, that connects the lower jaw (by means of the symplectic and quadrate bones) to the otic (auditory) region of the head; it supports the opercle series of bones, permitting its lateral bending for the animal to breath. IO, interopercle. LA, lacrimal. LE, lateral ethmoid. MP, metapterygoid, a bone connecting the hyomandibular bone (and therefore the skull) to the upper jaw anteriorly, by means of the ectopterygoid and mesopterygoid. MSP, mesopterygoid. MX, the maxillary; together with the premaxillary they form the upper maxilla, or the upper margin of the mouth. NA, nasal. PA, the parietal, one on each side of the head. PMX, the premaxillary. PO, preopercle. PTO, pterotic, a bone of the auditory capsule or otic region. PTP, post-temporal, a bone of the pectoral girdle that unites the cranium with this girdle. Q, quadrate bone, that supports the lower jaw by means of a ball-and-socket articulation that then permits the opening and closing of the mouth. RA, retro-articular, the mandibular bone located most posteriorly; it helps the up and down flexing of the mandible, thus opening and closing mouth. S, a bone of the sphenoid region. SO, subopercle. SOC, supra-occipital, the median most posterior bone of the head, which forms what we may call in these animals the "nape" of the head. SOR, series of orbital bones. SP, symplectic, a bone connecting the hyomandibular bone (and therefore the auditory region of skull) to the quadrate (Q) bone (and therefore to the lower and upper jaws). UH, uro-hyal, which helps connect the tongue with the posterior ventral region of the body. OP, opercle bone, a large bony plaque, which dislocates laterally, permitting the movement of water to the outside of the body, collectively working with the preopercle, interopercle and subopercle. The opercle articulates with the remaining head by means of a long, oblique bone, called hyomandibular, HM. For further information, see text.

MODIFIED FROM DEAN (1895) BASED ON GREGORY (1933), HARDER (1975: 12–13), LAGLER *ET AL.* (1962: 64), MOYLE AND CECH (1996: 23), WEITZMAN (1962) AND ON OUR OWN INTERPRETATIONS. IMAGE IN THE PUBLIC DOMAIN

The skeleton presented in the figures and described here is similar in most teleosts and not very different from the skeleton present in basal sarcopterygians. There is much variation among taxa. For example, sea horses (Teleostei) position themselves more vertically in relation to the sea bottom, so that anterior in these animals corresponds to ventral in most aquatic teleosts and sarcopterygians. Vertebrates associated with the transition to land position themselves horizontally in relation to water bodies, so these variations in body anatomy, and their corresponding nomenclature, will not be covered here.

In chondrichthyans, the upper maxilla (upper jaw) and lower jaw are formed by cartilage. The neurocranium is basically a continuous cartilaginous block. Vertebrae are also of cartilage. Each individual cartilaginous element may be strongly calcified, and quite hard, depending on the species. But as chondrichthyans have not made the transition to land, they will not be further discussed.

The skeletal elements provided above are much more detailed and more complex than those analyzed in Chapter 5. Characters effectively explored in this book are (1) the presence of a single proximal skeletal element in the paired locomotor appendages; (2) the pelvic girdle conspicuously developed; (3) the presence and development of the pre- and post-zygapophyses (described and

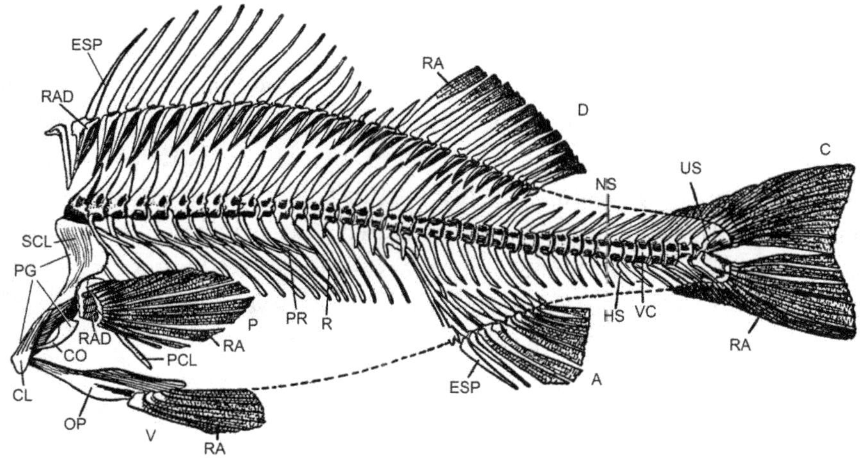

FIGURE 1-7 Post-cranial skeleton of *Perca fluviatilis,* in lateral view. Dorsal region upward. A, anal fin. C, caudal fin. VC, vertebral centrum. CL, cleithrum. CO, coracoid. D, dorsal fin, in this case divided into a part with spines and a part with rays. ESP, spine in the fin. HS, haemal arch and haemalspine. NS, neural arch and neural spine, which emerge from vertebral centrum and protect the hollow dorsal nerve chord in vertebrates. OP, pelvic bone. P, pectoral fin. PCL, post-cleithrum. PG, pectoral or scapular girdle, with its several subunits. PR, dorsal rib or epi-pleural rib. R, rib. RA, fin ray. RAD, radial, a skeletal element that supports the spines and rays of the fins. SCL, supracleithrum. US, urostyle. V, ventral or pelvic fin
MODIFIED FROM DEAN (1895), BASED ON WEITZMAN (1962). IMAGE IN THE PUBLIC DOMAIN

illustrated in Chapter 5); (4) the overall skeletal structure of the feet (described succinctly but in sufficient detail in Chapter 5); (5) the skeletal structure of the legs (also described succinctly, but appropriately for our argumentation, in Chapter 5); (6) the opercular bones; (7) the absence of an extrascapular bone, that belongs to the pectoral girdle; (8) the pleural ribs (R in *fig.* 1-7), and (9) the notochord (described, discussed, and illustrated in Chapter 5).

The skeletal characters above were described and illustrated for a representative of Teleostei, which belongs to the Actinopterygii. The nomenclature is similar and easily correlated to their homologues in their sister group, the Sarcopterygii, the group that effectively became terrestrial.

Although our book is quite simplified in anatomical and taxonomical coverage, and does not cover the rich fossil diversity available, this same simplification has permitted us to analyze a more representative diversity of taxa in our cladograms, centered at the level of the Osteichthyes as a whole. As a consequence, we have been able to observe that characters normally considered as having evolved for terrestrial life, in reality were already present long before the advent of terrestriality in vertebrates. This approach provided us with a general view, and an in-depth view, that permitted us to develop several new arguments that are contrary to the adaptationist view of the evolution of terrestriality. Therefore, we changed the focus from ecology to zoology, and from a restricted time scale to a more extended period of time. Our historical account is not restricted to geological events, but focused on zoological events (on anatomy and physiology, which are "internal" factors of organisms not constrained by external, ecological events). We thus valued the zoological attributes (characters, features) of the animals to a further extent than previous authors. They simply followed Darwin, who focused on natural selection as the main factor in evolution. Our results, expressed in this book, were not predominantly selective. We obtained this insight by valuing the attributes of the animals themselves, instead of the environment.

As a consequence of their prevailing adaptationist view (in which only natural selection is viewed as promoting evolution), most books dealing with the conquest of land by vertebrates deal only with sarcopterygians, and thus restrict their arguments to the group most intimately related to the terrestrial vertebrates. For example, Clack (2002, 2012), Laurin (2010), McGhee (2013), and Schoch (2014) provide excellent accounts of the evolutionary event in focus, but they have an adaptationist bias. This view does not represent a preconceived view of the authors, considering that they do not even consider discussing alternative views. In fact, they do not mention alternative possibilities, as if such other points of view had never been published. In fact, these authors apparently are not aware of the existence of other possibilities, and work under the adaptationist view as the only possibility available to them for the evolutionary process. This is possibly due to a historical point of view, that generated a sequence of researchers having the same outlook of the transition from water to land. This had become the view of their teachers and continued to be the view of advisors of graduate courses in the field of paleontology. The common practice in the qualification of a new professional in this field includes the description of new fossil material, that is then contextualized in an evolutionarily scheme, and then explained in adaptationist terms, in increasing detail. Therefore, it seems that only outside researchers, not involved in this historically constrained process of human professional qualification, could make alternative propositions. Because the Sarcopterygii represent a monophyletic group, these books, and in fact the almost totality of articles on the subject, see no reason to include taxa and characters occurring outside the sarcopterygians. We will see how useful the consideration of outgroups becomes along this book.

The importance of the skeleton was dealt with quite incisively by Hildebrand (1982). The internal, jointed skeletal system of vertebrates is unique in the animal kingdom and is the most important of all organ systems in the study of vertebrate morphology (Hildebrand 1982: 117). It is conservative enough in general pattern to show the broad outlines of vertebrate phylogeny: Homologous bones and evolutionary trends are readily demonstrated in skeletons, characterizing the successive major taxa (Hildebrand 1982: 117). Wide variations have always been superimposed on the gradually evolving general pattern, and the skeleton has been sufficiently plastic to respond to the particular habits of the various animals (Hildebrand 1982: 117). Therefore, it also provides reliable information about the specific adaptations of vertebrates: Posture and locomotor adaptations are accurately revealed, and other adaptations are sometimes indicated (Hildebrand 1982: 117). This general outlook of Hildebrand (1982) will be shown to be entirely coherent for several skeletal elements considered in Chapter 5, even when we are not dealing with specific characters. For example, the overall body shape, such as a "depressiform body" (dorso-ventrally compressed), is displayed by several taxa dealt with in this book. A general view of the fish skeleton will be helpful for the reader. Our general account of the skeleton of *Perca fluviatilis*, an actinopterygian teleost "fish", enables the reader to grasp our character-by-character account that will be followed more effectively.

For the beginning reader, Hildebrand (1982: 117) clarifies that, because of its hardness and durability, the skeleton (including the teeth) becomes fossilized relatively often. This contributes to virtually all (though not quite all) our knowledge of past vertebrate life. Paleontology can be regarded as the comparative anatomy and morphology of extinct animals. For Hildebrand

(1982: 117), no other Science has contributed so much to our knowledge of vertebrate evolution, and it is, of course, limited almost entirely to the study of hard parts.

Osteichthyes include Actinopterygii and Sarcopterygii. Both are characterized by the presence of endochondral bone in the endoskeleton (Janvier 1996: 65). Endochondral bones are not only ossified superficially (perichondrally), but also internally (endochondrally), in the form of what is usually called spongy bone. The endoskeleton is initially made up of cartilage. This tissue occurs in all craniates and appears to emerge first in ontogeny (Janvier 1996: 53). It is located internally in the body ("éndon" is a Greek word for "inside", "in the inner part of") and at the inner-side of the exoskeleton or dermal skeleton (Hildebrand 1982: 118, Janvier 1996: 53). The tissue may transform into bone in different ways: either the superficial tissues lining the cartilage produce bone (perichondral bone) or the cartilage itself is invaded by bone (endochondral bone) (Janvier 1996: 53–54). Both processes can occur simultaneously, but as only perichondral bone is known in jawless vertebrates, such as osteostracans or galeaspids, this fact suggests that it occurred earlier than endochondral bone (Janvier 1996: 54). There are many other characters unique to extant osteichthyans, according to Janvier (1996: 65–66). For us it is of direct interest that the fin-rays are covered by a particular type of elongated scales, called the lepidotrichiae.

As we presented in this book, for the osteichthyan skeleton based on an actinopterygian representative, the question is: are there important differences between them and the sarcopterygian skeletons, which are directly related to tetrapods? According to Janvier (1996), many differences can be pointed out. Janvier (1996) distinguishes four major and rather stable groups of dermal bones. The dermal bones would be the ones that belong to the endoskeleton, but which are not previourly formed in cartilage, such as some bones of the braincase of recent bony fishes (Janvier 1996: 54). (Therefore, the author distinguishes "dermal bones" from "membrane bones", which in turn are formed by the ossification of membranous connective tissue).

The maxillary (= maxilla in Janvier, 1996) is presumably lost in coelacanths and dipnoans (Janvier 1996: 67). Infradentaries or splenials emerged in sarcopterygians (Janvier 1996: 67).

The bones of the skull-roof consist essentially of paired large bones; from back to front in the skull are: postparietals (parietals in our figures), parietals (= frontals in *figs*. 1-5 and 1-6) and nasals (Janvier 1996: 67). Surrounding these there are a variable number of smaller bones, with names fluctuating as a function of theories on their homology, fusion, or loss (Janvier 1996: 67). Therefore, it is not possible for us to address completely the issue of nomenclatural diversity in our book, even using various independent lineage representatives of osteichthyan fish-like animals. Janvier (1996: 67) ends his exposition on the skull-roof by stating that posteriorly there may be an emergence of additional bones, such as the modified scales.

Regarding the cheek bones, Janvier (1996: 67) states that these include the circumorbital series (including supraorbitals, suborbitals, postorbitals, infraorbitals; the antorbitals are part of the anteriorly mentioned skull-roof), and a variable number of larger bones, the preopercular, squamosal, jugal and quadratojugal. Judging from the figures for the porolepiform †*Porolepis* (Janvier 1996: 202, *fig.* 4-77), the lacrimal may be the infraorbital 1 and/or 2 of teleost nomenclature; the jugal may be an antero-posteriorly elongated infraorbital 3 of teleosts; the lower, upper and accessory squamosals may have been detached from the posterior end of infraorbital 3 and/or 4. Also from this figure, it is possible to realize that the preopercle may be divided, at least in this genus. The supraoccipital is called the median extrascapular. This example shows that each sarcopterygian lineage may have not only a skeletal nomenclature of itself but new, different bones, with their respective meaning in the evolution of the group, and we do not gain much from entering into all these details. We would simply like to clarify that each species, each taxon, has its own group of evolutionary innovations (novelties). For reasons like this we find it unproductive to go into complex accounts of differences displayed by many subgroups of Sarcopterygii. The shoulder girdle does not seem to present remarkable differences able to cause any discomfort for the reader attempting to understand our arguments.

Janvier (1996: 67) considers that the methods of assessing homologies between dermal bones in different osteichthyan groups have been vigorously debated. For the aims of this book, only some generalized skeletal features are of interest for our argumentation and we will not go further in discussing feature details or in considering taxonomic variations.

CHAPTER 2

Adaptive Scenarios

An introduction to the theme of adaptive scenarios is briefly presented here. By adaptive scenarios we mean the adaptation of organisms to an environment which changes through geological time. This environment is supposed to select, through many generations, the variations which are advantageous for the differential transmission of DNA from one generation to the next. Several hypotheses have been erected by researchers who try to understand the evolutionary transition of vertebrates from water to land through adaptive scenarios. But other researchers which also have this point of view of evolution have published critical comments on these hypotheses. The main aim of this Chapter is to present a long sequence of proposals and their rebuttals already made in pertinent literature, avoiding entering our own criticisms, for now (except in one case).

It is not our intention, in this book, to simply criticize those that focus their research and interpretations on adaptive scenarios. Highly qualified professionals strive to understand the ecological context in which fossil organisms lived. The integrated knowledge of the fossil biota that lived dozens or hundreds of millions of years ago is allied to the focus on natural selection, considered to represent the main source of biological evolution (actually the single one, as these researchers do not mention others). Such a frame of mind produces evolutionary theories based on the geological and ecological evolution of our planet before, during, and after the attainment of terrestriality in the Tetrapoda. All these researchers have expended an enormous effort to understand the origin of terrestriality in tetrapods from an adaptationist point of view, centered on adaptive scenarios. This becomes problematic. In the words of Clack (2012: 140), most ideas can be countered by objections or alternative suggestions, or are difficult or impossible to test.

Our approach in this book is furthermore not to imply that natural selection has not influenced the origin of terrestriality during the evolution of tetrapods. Simply stated, we propose that macroevolutionary events were much more important than has previously been perceived. We intend to show that such macroevolutionary events explain more adequately various issues discussed in the adaptationist program, but that prove so difficult to clarify under this point of view. We further conclude that macroevolutionary events represent the dominant factors involved in the origin of terrestriality. This alternative view has emerged from our research and will be detailed in subsequent Chapters.

 | DOI:10.1163/9789004684911_003

We discuss the terms "microevolution" and "macroevolution" in Chapter 3 in more detail. At this moment, it is necessary to explain that microevolution is an evolutionary process resulting directly from natural selection, which continually refines characters. Macroevolution includes evolutionary paths that occur more abruptly, and which do not depend on evolutionary continuities.

Geological time is so long and complex that it has been necessary to devise a geological time scale. The planet Earth is about 4.5 billion years old, of which about the last 600 million years contains the fossil record of animals. This last time period is known as the Phanerozoic. It is divided into three Eras: Paleozoic, Mesozoic (better known as the era of reptiles, when the dinosaurs dominated our planet), and Cenozoic (after the extinction of the dinosaurs, excepting their bird descendants), which includes the present time. Here we are interested in the Paleozoic Era, in which the transition of vertebrates from water to land occurred. This era is divided into periods, the first of which is called the Cambrian (up to about 490 Myr). The Ordovician follows (up to approximately 443 Myr), then the Silurian (to 416 Myr), the Devonian (until about 360 Myr), the Carboniferous (to approximately 300 Myr), and finally the Permian (until about 250 Myr). The Devonian (as the other periods) was divided into Ages, and these into Stages. Paleontologists specializing on the origin of the land vertebrates deal extensively with these subdivisions. Table 2-1 shows the evolutionary time scale, in which the several partitions of time of interest for this book are illustrated in a simplified way (see legend for further details). We also indicate the approximate geological moment in which the so-called morphological apomorphies appeared. These will be central to our argument, directly in *fig.* 5-1. Furthermore, several apomorphies appeared before the Devonian, while others appeared close to the limits between the Devonian and the Carboniferous, at least on the basis of the currently known fossil record. According to Niedzwiedzki *et al.* (2010), some uncertainty remains regarding the timing of taxa at the level of genera. Thus, presenting exact original datings and more precise geological scaling would not resolve the problem of precision of these data. For this reason, in Chapter 6, we provide more extended time periods. We wish to validate our reasoning, until new fossil records come to alter known dates of apomorphies.

The importance given to the ecological environment for the appearance of terrestriality may be deduced by the presentation of a Chapter dealing exclusively with the Devonian period in both books of Clack (2002, 2012). Details on geological fluctuations in oxygen and carbon levels are presented, leading the reader to believe, even when not explicitly stated, that increases and decreases in parameters such as these may have been important factors for the origins of terrestriality. Modern computational models use factors such as

TABLE 2-1 Geological scale of the Earth. Geological time is shown as progressing from below to above, in the table. The beginning of each Period is shown in average millions of years. The space occupied by each period in this table is not proportional to the real time and its duration. Some periods occupy more space in the table to accommodate information on taxa or events, in the right-hand column. Information on mammals and humans are provided only as reference points for the reader. †*Pederpes* is the oldest completely terrestrial vertebrate from the known fossil record. †*Panderichthys* and †*Tiktaalik* are vertebrates without feet more intimately related to recent Tetrapoda (vertebrates with feet). †*Acanthostega* is the oldest tetrapod for which the skeleton is almost completely known. †*Ichthyostega* is the tetrapod related to the origin of the Tetrapoda best known to the great public. Adapted, modified and greatly simplified from Newman (1923: 26–27, *figs*. 5-6), following models in Romer and Parsons (1977), Clack (2012) and Schoch (2014). Dates approximate, as in Gradstein *et al.* (2004 *ap*. Clack 2012).

Era	**Period (time in millions of years ago)**		**Events or taxa linked to theme of book**
CENOZOIC	1	QUATERNARY (present)	*Homo sapiens*
	65	TERTIARY	
MESOZOIC	145	CRETACEOUS	
	200	JURASSIC	
	251	TRIASSIC	First mammals
PALEOZÓIC	300	PERMIAN	
	318	Upper CARBONIFEROUS	
	360	Lower CARBONIFEROUS	*Crassigyrinus,* *Whatcheeria* *Pederpes*

TABLE 2-1 Geological scale of the Earth. Geological time is shown as progressing. (*cont.*)

Era	Period (time in millions of years ago)		Events or taxa linked to theme of book
	416	DEVONIAN	*Acanthostega, Ichthyostega* First tetrapods *Panderichthys, Tiktaalik* First lung fishes
	444	SILURIAN	First bony fishes
	488	ORDOVICIAN	
	542	CAMBRIAN	

calcium silicates, magnesium and pyrite, degradation of organic matter, degree of exposure of coastal lands and variabilities in degradation of vulcanic rocks, in order to obtain evolutionary estimates. Totally different results appear in publications along the years. When previously oxygen levels appeared to vary from 19% at the beginning of the Devonian to 33% at about the middle of this period, now the variation seems to have been from 22% to 13%, during the same period (Clack 2012: 101–102, *fig.* 4-1). This is about the opposite context regarding oxygen levels. Contradictory data do not seem to challenge adaptationist interpretations. *Ad hoc* arguments, propositions or hypotheses are statements having the single objective of legitimating or defending a particular theory, not the result of an objective understanding of reality (Houaiss and Villar, 2009: 49). We must thus be alert for possible *ad hoc* explanations provided in order to justify specific adaptationist positions.

Some researchers believe that life on dry land began in the Lower Devonian (*e.g.* Clack 2012: 102), while others state that plants were already firmly established in coastal areas and the margins of rivers throughout our planet in the

Silurian (*e.g.* McGhee 2013). Clack (2012: 106) recognizes that complex plant communities are known from the end of the Silurian (Tomescu and Rothwell 2006 *ap.* Clack 2012). By the end of the Silurian, true vascular plants had evolved, with transporting systems for the conduction of water and for gaseous exchanges. Clack (2012: 107) indicates that in the Lower Devonian plants were more complex, being more ramified and having structures similar to leaves, some attaining a height of 2 m. There are no records of vertebrates living on land before the Carboniferous. The Devonian spanned around 57 million years. This is the available time span for selective pressures to have acted on favorable adaptations for the progressive conquest of land. As there is already evidence for the presence of arthropods at the end of the Silurian (Jeram *et al.* 1990 *ap.* Clack 2012: 106), such an extended period of time, so far removed in the geological scale, seems to us to involve a disconnection between data and interpretation.

According to McGhee (2013: 65), the fossil record contains myriapod arthropods (centipedes and millipedes; Brusca and Brusca 2007) and arachnids on land during the middle of the Silurian. The evidence even indicates an explosive radiation rate of lineages of millipedes in the Silurian. There appears to be a distinct discrepancy in the colonization of land by plants and arthropods on the one side, and vertebrates on the other. Land vertebrates occur more than 50 million years later, maybe even 60 million years after plants and arthropods (a period of time nearly as long as the Cenozoic Era). Although there is no published method to correlate microevolution with geological time, it seems clear that the invasion of land by arthropods is correlated with that of plants. Despite the existence of favorable conditions for the occupation of land, vertebrates invaded land much later. This puts microevolution in doubt as representing the main evolutionary process. If natural selection (microevolution) were the process responsible for the conquest of land in vertebrates, one would expect that the selected character, or set of characters, would be quite diversified, and selected much more rapidly, given the evolutionary advantages they provide. Land would be conquered in a relatively short time span, such as occurred with terrestrial invertebrates following the previous occupation by plants. It would thus appear incoherent to associate a microevolutionary process with a time span of 50–60 million years. Ecological factors were totally favorable for this event at least 50 million years before the vertebrates actually conquered land.

Clack (2012: 98) appears to have pinpointed the beginning of the present focus on geological environmental scenarios for organic evolution. She refers to the New Synthesis, in which evolution is viewed in terms of adaptive radiation. Species appeared as taxa adapted to certain environments. Adaptations to particular environments are seen as more relevant than phylogenetic descent for certain animal characters. In such a context, it is to be expected that the research that followed on the origin of tetrapods would be focused on adaptive reasoning. As a result, several adaptationist hypotheses appeared. Some will be presented below, succinctly.

A Devonian environment studied in detail refers to the red sand layers, called the Old Red Sandstone. This particular environment had a large repercussion on theories for the appearance of land vertebrates. According to Clack (2012: 133), the known Devonian sediments were arid or desert-like, being represented by red sandstone. Initial theories proposed that arid conditions provoked the recession of lakes and rivers in which sarcopterygians ancestral to tetrapods lived. These animals thus ended up by becoming stranded in disappearing bodies of fresh water. Those having lungs and being capable of breathing atmospheric air were more capable of surviving. One problem with this idea, pointed out by Clack, is that many representatives of these sarcopterygians, with the appearance of fish, survived from the Devonian to the Carboniferous and even to the Permian. Currently, lungs are supposed to have evolved in an exclusive common ancestor of Sarcopterygii + Actinopterygii. They thus are an apomorphy of Osteichthyes, which comprise both of these two groups of vertebrates. Basal Actynopterygii would also be benefited from having lungs (*e.g.* Standen *et al.* 2014) and should have colonized land, a fact that the fossil evidence does not support. Some authors perceived that the red sandstones are not necessarily correlated with an arid climate. But this discussion is less relevant, because there may have been arid regions of red sandstone at the time of origin of terrestrial vertebrates. The biological/phylogenetic counter-argument is the most important, as we have seen above. Other arguments presented by Clack (2012: 134) are also less relevant than the biological argument: even if the deposits of red sandstone are arid regions, the ancestors of tetrapods may not necessarily have lived in them. Forested areas could have occurred around water bodies even in such arid environments. Presumably, the first tetrapods may have appeared before the end of the Devonian. Maybe they appeared even before the occurrence of the red sandstones. Thus, they may have originated in more humid environments. The ecological conditions in which the first tetrapods appeared are not well known. Whether they were correlated with environments subject to tidal variations, or depositions of red sandstones, have no consequences at all.

Low oxygen concentrations in the water and in the air, together with increases in temperature, which occurred in the Givetian-Frasnian, may have gradually modified environmental conditions (Clack 2012). However, these conditions became inverted by the end of the Devonian, as can be seen in Clack (2012: 102; *fig.* 4-2). Trying to associate all these variables with selective pressures for the conquest of land may become quite flippant. Tetrapods may have confronted all these situations simply using their lungs, present already before the Devonian. We may mention the basal actynopterygian *Polypterus,* capable of surviving even out of water (*e.g.* Standen *et al.* 2014). The basal teleost *Arapaima gigas,* the famous pirarucu, the largest freshwater fish in South America, breathes only atmospheric air through a structure in some ways intermediate between a lung and a swim bladder (Lauder and Liem 1983: 125). Thus, selective pressures in favor of terrestriality do not seem to occur. The adaptive scenario is difficult to sustain. Basically, the same type of criticism was made for the hypothesis of Romer (which we discuss immediately below). The existence of a lung would counter any dilemma much before the appearance of any advantage in switching to land.

For Romer (1966: 86) and his contemporaries, there must be a reason for vertebrates to emerge onto land. There should be a good evolutionary explanation for the development of land vertebrates. Romer (1966) argues that there was little food on land, but there were no aquatic enemies to avoid in the terrestrial realm. The aquatic vertebrates from which they descended were among the most powerful freshwater forms. He believed that the Devonian was a time of periodical dry seasons. When water pools became totally dry, the fish would be trapped in the mud and would die. But the amphibians, with their short and clumsy, yet effective appendages, would be able to crawl away until they found another water body, resuming their normal aquatic existence. For Romer, these animals would feed gradually on fish trapped in the mud, later on terrestrial insects, and finally on terrestrial plants. The largest carnivorous amphibians would be able to feed on smaller amphibians. Thus, a fully terrestrial fauna would finally become established. For some critics, however, animals capable of returning to water would not have incentives to remain on land. In fact, they would remain comfortable in aquatic environments. This idea of Romer was thus abandoned (*e.g.* Clack 2002, 2012). Another problem with this idea is that many representatives of these sarcopterygians, with the appearance of fish, survived from the Devonian to the Carboniferous and even to the Permian (Clack 2012: 133).

Notwithstanding the strength of these criticisms, several other proposals that were criticized with similar arguments were created. Carroll (1988: 166) considered that the availability of food on land (contrary to Romer) and the freedom from competition with other rhipidistians (sarcopterygians excepting coelacanths, lungfish and †porolepiforms [*vide fig.* 5-1]), favored the development of walking legs in tetrapods. Larval stages are perceived as smaller and, supposedly, more agile. They may more easily live at the margins of water bodies. They would even be able to crawl between water and land more easily.

However, the more basal members of anurans, salamanders and limbless/eyeless amphibians, and most living representatives of the last two groups, develop their eggs and juveniles inside the body of the mother (Clack 2012: 135). The critics to Romer's idea also apply to this hypothesis. Proposals in which paired locomotor appendages, still in the evolutionary stage of pectoral (anterior) or pelvic (posterior) fins, with or without the development of feet, could be used by the individual to crawl or walk among the aquatic vegetation can be criticized with similar arguments.

Clack (2012: 136) refers to "walking" in *Periophthalmus*. Harris (1960), on the other hand, does not refer to any kind of walking in *P. koelreuteri*. Six types of locomotion were characterized by him for this species, indicating that the pelvic fins are relatively short and serve mostly for body support rather than for giving a clearly delimited "step". In any case, the critics to Romer's hypothesis apply again. Another questioning is the same as the previous criticism made in this Chapter for the idea of Romer: Carroll *et al.* (2005) argue that the first tetrapods and some of their closest relatives (†elpistostegids sometimes referred to as †panderichthyids) were dorso-ventrally compressed in body form because shallow waters tend to have higher temperatures. This would elevate body temperature and, consequently, the metabolic rate. Once the body is warmed, the selective pressure to remain on land ceases. For Clack (2012: 137), one of the most common reasons for aquatic vertebrates to leave water is the obtention of food from terrestrial or semi-terrestrial sources. According to her, it has been demonstrated that fishes (= teleosts, a subgroup of Actinopterygii) that feed on land and remain there for certain periods of time enhance their digestive rates. Digestive enzymes and gastric acids probably function best biochemically at higher temperatures, such as those promoted by the sun. Clack says that fish that feed in the water rise to land in order to benefit their digestive rate. High environmental temperature would be the key factor. However, once such adaptations are attained, the organisms become adapted to their aquatic environments and lack further selective pressures to conquer the terrestrial environment. That the animal leaves the water in order to become warm is not an entirely implausible proposal, but suffers from the same criticism as Romer's hypothesis, and several other hypotheses could be envisioned. Byrne *et al.* (2018) hypothesize that the presence of strongly modulated tides may have been a significant driver in the transition. A large tide just after the first or third quarters of the moon, when there is the least difference between high and low water in spring, would have created a large intertidal zone according to them, thus stranding sarcopterygians for several days. They reached the idea of the existence of strong modulated tides by producing tidal simulations of the Siluro-Devonian period using paleogeographic plate reconstructions for two time slices prior to the Polish trackway date. For us, this is not different from Romer's idea of aquatic vertebrates being stranded in the dry season. Clack (2012: 136–137) noted that teleosts remain for some time on firm land for a series of reasons and circumstances, as accounted by Sayer and Davenport (1991), suffering from the same criticisms as Romer's hypothesis, among other possible criticisms.

The same is true regarding temperature. Some periophthalms only leave the water when the temperature is sufficiently high. Some cloud fish only leave the water when the water temperature is too low (Clack 2012: 137). But these adaptations seem to be sufficient for the survival of the species belonging to these groups. These species, then, are well adapted to their aquatic environments, there remaining no further evolutionary pressures for them to leave the water or to conquer land.

The presence of a labyrinth organ in some teleosts, drinking water from the surface, gulping air to pass it through the gut in others, developing lungs in osteichthyans, were all supposedly selected by the same selective force(s). Yet only one taxon developed lungs (that is, lungs emerged only once, not in every aquatic vertebrate group). Therefore, the same selective pressure selects different variations in different evolutionary lineages, so that not a single and the same feature will always emerge. Because of this, only a single subgroup of the gnathostomes, and only a single subgroup of the sarcopterygians, originated the terrestrial forms. Dipnoi, a subgroup of sarcopterygians (*fig.* 5-1), survives occasional droughts to the present day, and has never left the water or originated a group of terrestrial vertebrates. Thus, in South America and Africa, the genera *Lepidosiren* and *Protopterus,* respectively (living genera of Dipnoi, *e.g. fig.* 2-1), undergoe periods of droughts (with a strong reduction in their activities, including their metabolism) during the dry season. They have functional lungs and are frequently called "lungfish". Just before the river water dries out, they burrow into the bottom mud, leaving an opening to the surface. They remain for month within their burrows, breathing atmospheric air and consuming their energetic reserve tissues, mainly fat. Thus, drought events are an *ad hoc* argument for the conquest of land. This argument was created to explain a microevolutionary process that would have selected organisms to endure such an environmental stress. They supposedly have led to the origin of terrestrial vertebrates, but Dipnoi survived this environmental stress differently: they enhanced their capacity to remain in their original aquatic habitat, instead of leaving it and invading land.

Copella arnoldi (Géry 1977: 143, 146) illustrates the fact that at least some teleosts spawn out of the water. As Clack (2012: 138) observes, they spawn at the interface of the water, near its surface. In this way the lower tide exposes the eggs to air and to daylight. Their temperature increases metabolism and,

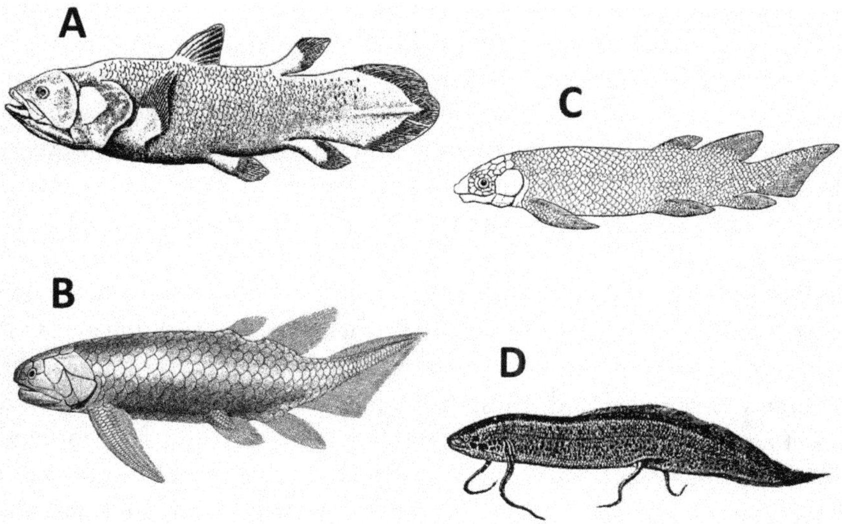

FIGURE 2-1 Non-rhipidistian sarcopterigians. Lateral views. A – *Latimeria*, (Coelacanthiformes). B to D are Dipnomorpha. B – †*Holoptychius* (Porolepiformes). C – †*Dipterus valenciennesi* (Dipnoi). D – *Protopterus annectans* (Dipnoi), note the filiform paired fins (pectoral and pelvic) and the fusion of the dorsal, caudal and anal fins.
CREDITS ARE AS FOLLOWS. A, FROM HTTPS://COMMONS.WIKIMEDIA.ORG/WIKI/CATEGORY:LATIMERIA_CHALUMNAE#/MEDIA/FILE:COELACANTH-BGIU.PNG ATTRIBUTED TO BOGDAN; SEE ALSO HTTPS://COMMONS.WIKIMEDIA.ORG/WIKI/FILE:COELACANTH.PNG BY ROBBIE CADA (MAY 21TH, 2018). B, FROM HTTPS://COMMONS.WIKIMEDIA.ORG/WIKI/FILE:HOLOPTYCHIUS_FLEMINGI.JPG (LANKESTER AND RIDEWOOD 1908) HTTPS://ARCHIVE.ORG/STREAM/GUIDETOGALLERYOF08BRIT#PAGE/64/MODE/2UP; JUNE 25TH, 2018). C, MODIFIED FROM GOODRICH (1909). D, FROM KINGSLEY (1889). ALL IMAGES ARE IN THE PUBLIC DOMAIN

thus, abbreviates the period of embryo development. As the fertilized eggs of *C. arnoldi* remain out of the water and are able to sustain the effects of the sun, at least during some minutes during the day, it is possible that they benefit from temperature increase and accelerated embryonic metabolism. In the case of this species, the location of eggs out of the water reduces their predation. Terrestrial animals do not normally feed on fish ova, probably because they do not recognize such ova as food. This reproductive strategy may thus be advantageous for this species. Nonetheless, here ends the evolutionary pressures in the direction of a conquest of land.

Fossil trails of "scorpions" belonging to the group Eurypteridae were found in Australian deposits dating 420 million years ago (Upper Silurian period). These tracks belong to animals of large size, maybe attaining 2 m in length.

They are considered to represent direct evidence of arthropods that emerged from water (McNamara and Selden 1993). These fossils indicate the presence of predators capable of intimidating terrestrial vertebrates, and these predators were already present on land dozens of millions of years before the first known truly terrestrial vertebrate fossil (maybe as much as a hundred million years before, by the present fossil record). This shows that the absence of potential predators from land, in contrast to their presence in aquatic environments, must be reviewed.

With this information of giant eurypterid tracks in the Upper Silurian, we may conclude that food was also present for them. Clearly, this is the food that several researchers have argued as representing the selective pressure for the vertebrates to conquer land. Thus, there already were invertebrates on land at least 420 Myr ago (Upper Silurian). These could represent food for land vertebrates. However, the fossil record of terrestrial vertebrates is less than 360 million years old. By the adaptationist point of view, selective pressures provoked by the presence of food items on land could have led vertebrates toward land about 60 million years before the known conquest of land by vertebrates, considered to have occurred in the beginning of the Carboniferous. Therefore, the adaptationist viewpoint is challenged by data known for a long time.

McNamara and Selden (1993) point out that there was a large diversity of invertebrates living on land, including the 420 Myr old, 2 m-length eurypterid "scorpions". On the other hand, they consider the large diversity of predatory invertebrates in the seas of that time (McNamara and Selden 1993: 25). Predatory arthropods living in the seas, rivers and lakes include these same eurypterid "scorpions", as well as diverse mollusks. They point out that nevertheless an "explosion" of fish lineages, in these same environments, occurred 420–400 million years ago, many of them being predators (McNamara and Selden 1993: 26). The increase in aquatic vertebrate diversity would increase competition in water. This would represent a selective pressure for the conquest of land (McNamara and Selden 1993: 26). But a criticism may be put forward based on the fact that only a single group of vertebrates conquered land (Triques and Christoffersen 2009). If selective pressures had acted for vertebrates to leave water, several aquatic groups could have originated the terrestrial vertebrates, forming a polyphyletic assemblage. Yet a single species of vertebrate originated a large, unique group of terrestrial vertebrates, the monophyletic Neotetrapoda, or the crown group of the Tetrapoda, as we will see in following Chapters.

Robust tree trunks are known to have evolved independently twice in Devonian: in the Middle Devonian and in Upper Devonian (stages Givetian-Frasnian) (Algeo *et al.* 1995 *ap.* Clack 2012) but truly terrestrial vertebrates are only known from the Carboniferous (Clack 2002, 2012). The establishment of a causal correlation between plant vegetation and vertebrate terrestriality may therefore be questioned.

Finally, there is a hypothesis that appears to represent a lateral branch to the original thinking of Romer, which we discussed previously. This is the idea that aquatic vertebrates could become stranded on the beach during low tides. This would represent a selective pressure for other vertebrates to leave the water to eat the stranded animals, culminating in the eventual conquest of land. This hypothesis stumbles on the fact that, with the high tide, these same dead animals would become available for the aquatic carnivorous vertebrates, eliminating the supposed selective pressure by an interval of hours. Another criticism is that walking legs would have to be present and functional, so that another adaptive previous scenario is necessary to explain this adaptation. And then, it would no longer be necessary to invoke moribund animals lying along the beach to explain the conquest of land.

A similar *ad hoc* hypothesis was recently resubmitted. Byrne *et al.* (2018) recognize that vertebrates have made a sea-to-land transition only once (as already stated by Triques and Christoffersen [2009], not cited by them). It must be assumed that this transition was prior to 395 Myr, after the discovery of stem-tetrapod trackways in Poland. They state that the "biological perspectives" suggest that it was a gradual progression.

An interesting argument would be that any *ad hoc* hypothesis may explain part of a selective pressure that may have acted *in combination* with an exaptation. However, this argument fails in that the exaptation requires the existence of a complete feature to be exapted to *prior to* the event of emergence of a selective pressure toward the new environment or life style. Therefore, the evolutionary process we are considering is exaptive in nature. Without the exaptation event, there is no possibility for the subsequent selective pressure. In this case, selective forces would be of secondary importance in time and with regards to evolutionary mechanics.

At this point, we stress that there is no direct passage from an adaptive series generating a character (the primary adaptation) to another adaptive series gradually changing the use of the character (the secondary adaptation). What happens between them is an exaptation, an abrupt occurrence. This evolutionary event occurs because the structure is already present, complete, and then it has its use changed, which is the definition of exaptation. That is, an abrupt change of use occurs. We call this exaptation. After the evolutionary event of exaptation, another adaptive series emerges (the secondary adaptation). Exaptation triggers the second adaptive series. This second series was not triggered by the first adaptive series, as usually inferred, but rather by the abrupt change in use, the exaptation. Therefore, exaptation does not happen simultaneously with the second adaptive series, but rather before the secondary adaptation. Such secondary adaptations are therefore secondary both in terms of evolutionary momentum and in terms of character aptitude.

All such adaptive hypotheses, published for decades, were crucial for maintaining our interest in, and determining the centrality of, the issue of terrestriality in Tetrapoda. Had previous authors not tried to contribute to this major enigma in evolution, we would not have started our own analyses on the issue (many more hypotheses exist but have not been mentioned here). Our aim has been to demonstrate inconsistencies that, as a result, permit us to provide our own conjectures on this interesting theme. We consider all these previous attempts to clarify the subject of terrestrial life in tetrapods scientifically relevant. We also point out that a critical attitude toward previous ideas always represents, in the long run, a healthy stance for the advancement of Science.

CHAPTER 3

Adaptationism, Microevolution, Macroevolution, and Exaptation

The theory of the origin of species by natural selection, or the "preservation" of "races" favored in the struggle of life, beyond being the title of Charles Robert Darwin's most important published book (Darwin 1859), was also the description of a revolutionary idea that freed human thinking from the religious dogma of the time. It permitted much more fluidity to the inquisitive mind, and for human life. Darwin associated variability, naturally perceived by biologists, ranchers, and other professionals of the time, to the process of natural selection. This undirected, "blind" evolutionary process, made the available theological or spiritual explanations of vital processes unnecessary (*e.g.* Futuyma 1995). The importance of the changes that these ideas provoked cannot be minimized. The liberty to think permitted scientific and technological development and influenced most aspects of human knowledge.

The two main theses claimed in Darwin's (1859) book are (1) that organisms are the product of a history of descent with modification from their common ancestors, and (2) that the main mechanism of evolution is natural selection of the inherited variations. Hereditary variations finally influenced biological research lines dealing with the evolutionary history of organisms and permitted the elucidation of evolutionary mechanisms (Futuyma 1995: 8–9). However, Darwin himself, with his extraordinary insight, noted that probably other evolutionary mechanisms also lead to organic evolution, as we mention with more details later. Genetic drift is a known evolutionary mechanism which is independent of natural selection, for example. Perhaps because natural selection has and will continue to have such a major biological and social impact, the adaptation of characters to their environment by way of natural selection has become the working focus of the most prominent evolutionary biologists. Other biological processes ended by being less explored. Natural selection occurs greatly (or totally) in response to the environment. It is expected that environmental changes select different characters (or character variations) through time as a function of the new geological conditions, including environmental and vegetational factors, favoring certain variants in detriment of others. Thus, the process of evolution by natural selection is perceived as being very detailed in time, occurring progressively along each generation of organisms. The perception of the small-scale evolution thus produced led to

 | DOI:10.1163/9789004684911_004

denominations such as "microevolution" for the level of changes in the gene frequencies of a given population.

Natural selection selects favorable variations for a particular environment. The selected variant feature may be and has been called adaptation to the environment. As a consequence of this perception, researchers tried to discover what an adaptation represents (in a case-to-case fashion) and which selective forces act to select the variations of a character. Adaptationism, then, is a research strategy that aims to identify adaptations and the selective forces that direct evolution in past environments (Andrews *et al.* 2002). However, since the middle 70s, paleontologist Stephen J. Gould and geneticist Richard Lewontin have been critical of adaptationism, particularly as applied to understand human behavior and cognition (Andrews *et al.* 2002). The criticism refers to the acceptance of stories (hypotheses) without sufficient empirical evidence (Andrews *et al.* 2002). In particular, Gould, Lewontin, and their colleagues argue that adaptationism frequently uses inappropriate patterns of evidence to identify adaptations and their functions. They often fail to consider alternative hypotheses for the idea of adaptation (Andrews *et al.* 2002). In response, Blackburn (2002) proposed a method to distinguish exaptation from adaptation, suggesting the use of cladograms as a basis for this process. We detail this method in Chapter 5. Gould and Vrba (1982) define exaptation as a structure (feature, trait, terms that in our book are treated as functional synonyms) that currently enhances the success of an organism in its environment, but which has not arisen by natural selection for its present function. These authors propose that adaptation, as suggested by Darwin, is a concept restricted to structures built by natural selection for their present function. Gould and Vrba clarify that the etymology of "adaptation", *ad* + *aptus*, that is, in direction of the adjustment, refers to a particular function. When a man-made tool is altered for a different use, it is consciously altered so as to function well for its new planned task. The authors argue further that, when creationists speak of adaptation, they refer to God's wisdom in projecting organisms for specific roles. Another example provided by them is of physiologists, when they say that larger lungs in Andean people are adaptations for local climates, a change directed toward a better functioning of these organs. Gould and Vrba (1982) note that in these examples, "adaptation" is defined specifically for the function it exerts. On the other hand, in biological evolution we find different uses for adaptations, and a possible conflict in concepts called adaptations (Gould and Vrba, 1982). The first use is consistent with the vernacular uses mentioned above. A feature is only considered as an adaptation if it was generated for its current function by natural selection. The second use defines adaptation in a direct or static way, as any structure that enhances present adaptation in relation to the historical origin of the structure. Probably as a consequence, Gould and Vrba (1982) note that another misunderstanding may occur. This refers to the use of adaptation both as an evolutionary process and as the state of a structure or of a trait. These authors restrict their discussion to the "state of being" of the structure, that is, to structures that contribute to fitness. Gould and Vrba (1982: 5) refer to the classic book of Williams (1966), indicating that the author had already recognized this problem. Williams restricts the term adaptation to the first use, or to the most restricted notion. Williams (1966: 6 *ap.* Gould and Vrba 1982: 5) says that we should speak of adaptation only when we are able to attribute the origin and perfection of this structure (or trait) to a long period of selection for the enhanced efficiency of a particular function. According to Gould and Vrba (1982: 5), Williams also argues that we should distinguish adaptations and their functions from their fortuitous (chance, accidental) effects. Williams uses "effect" in the vernacular sense – something caused or produced as a result or consequence. The concept of "effect" of Williams may be applied to a character, to its use, or to a potential process that appears as a consequence of a true adaptation. Fortuitous effects (by chance) always have the connotation of a consequence that follows "accidentally" and does not appear directly by construction (or origin) by natural selection. According to Gould and Vrba (1982), Williams usually invokes the term "effect" to designate the function of a useful character that was not generated by natural selection for its present (current) use – and they then confirm that they follow this restriction of Williams. Williams also recognized a certain haggling regarding the significance of the term adaptation, which is encouraged by imperfections of terminology (Williams 1966: 8, *ap.* Gould and Vrba 1982: 5). Thus, an imprecision over the central terms of evolutionary theory has been identified. This confusion arises because of the lack of a term connecting form with fitness. According to Williams (1966), authors term any structures that promote fitness, and that were developed by natural selection for their present function, as adaptations (criterium of historical genesis).

Gould and Vrba (1982: 6) call characters that evolved for other functions (of even for no function or no use whatsoever) exaptations. Such exaptations were coopted for their present use only later. They are apt for their present use, thus *aptus*, but were not developed for such a use, and are thus not *ad aptus*, or did not reach their present aptness by fitness. Their fitness is due to other reasons, and they are apt (*aptus*) *by way of* (*ex*) their shape, or *ex aptus* (italics as in Gould and Vrba). To clarify, adaptations have a function, while exaptations have an effect. As an example, we suggest that the gain of hairs in mammals (or of a slightly larger group of Therapsida – *e.g.* Pough and Janis 2019) was a tactile adaptation (because of concerning innervation), that was later co-opted for thermal isolation (*e.g.* Straile 1960, 1969, Maderson 1972). Thermal isolation is fundamental for animals producing their own body heat (called endothermal animals), to avoid loss of heat to the environment at rates that could lead to death. Note that, if the hairs had not arisen for another (tactile) function, there would be no possibility of a new selection, now for covering the body to avoid heat loss to the environment, an effect of the previous existence of hairs. Thus, hairs are here considered exaptations for the thermal isolation of mammals. Note that there was also selection for increasing the body coverage by fur (and selection of finer hairs, which cover the body more closely and more densely). This is not the same process of selection that led to the origin of the structure (hair) but represents a second round of selection that led to a new function for this structure (now said to be an effect). The important point here is to note the role of chance in the process that permitted thermal isolation in mammals: if hairs had not appeared previously as tactile structures, this mode of thermal isolation, so practical in terrestrial environments, would never have been able to appear. Thus, we do not see a continuous process of microevolution of structures for the production of thermal isolation, but rather we detect a macroevolutionary event: the abrupt change of use of a feature that otherwise would never emerge (for thermal regulation). A co-optation between a previously existing structure, with a previous function, and a new function, or a new "effect", as Gould and Vrba (*op. cit.*) prefer, is seen to occur. If a feature had not appeared for one use, it would not be present for a secondary use, which is a chance event. DNA variation could have been different, offering a different variation positively selectable by natural selection, generating another useful feature for this original selective pressure, but that could not serve a secondary use. In other words, the original selective pressure could have selected a feature not useful for the secondary use. So, this is the chance event. For instance, selective pressures for bottom dwelling generated the cracker-shaped †*Phyllolepis* †placoderm, the banjo catfishes (Actinopterygii, Teleostei, Ostariophysii, Siluriformes, Aspredinidae), the batfishes (Actinopterygii, Teleostei, Paracanthopterygii, Lophiiformes, Ogcocephalidae, *fig.* 3-1), the angel shark and rays (Chodrichthyes), but only the widespread pectoral fin, continuous to the sides of the head in rays, permitted propulsion by pectoral fin undulation. This finally led to a planktivorous sea level existence of manta rays, a clear exaptation of the depressiform body shape of rays. The other depressiform bottom dwellers did not generate surface-living vertebrates, so planktivory represents the chance event.

The history of any complex feature, structure, or trait, as Gould and Vrba (1982) conclude, quite probably includes a sequential mixture of primary adaptations, exaptations, and secondary adaptations. Any coopted structure

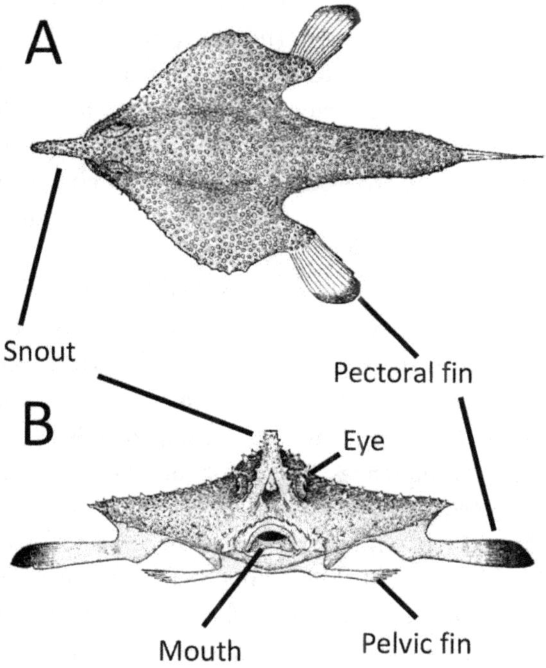

FIGURE 3-1
Batfish *Ogcocephalus vespertilio* in dorsal (A) and frontal (B) view
MODIFIED FROM GOODRICH (1909). IMAGES IN THE PUBLIC DOMAIN

(an exaptation) will probably not appear in a perfect way for the new effect and thus will develop secondary adaptations for its new role. In our example, hair was selected as a stiff sensory feature. After being exapted for body thermic insulation, a new set of selection or selections started and then led to minute changes toward the so-called guard (flexible and elongated) and wooly (flat in cross section) hairs, which are not stiff (Hildebrand 1982: 104). From a sensory function, the hairs were exapted for insulation by way of multiplication (Hildebrand 1982: 104) for the new use (an effect of its first emergence by selection). Even the "simple" hair, compared to the more complex tetrapod paired appendages, for instance, undergoes a "secondary adaptation", in the words of Gould and Vrba (1982: 12). As these authors state, the primary exaptations and secondary adaptations can, in principle, be distinguished.

In sum, to be considered an exaptation, the origin of the feature cannot be evolutionarily correlated with its current use. The feature's origin may have been by a selective pressure having no relation to its current use, or its origin may have been by genetic drift or even as a by-product of the origin of another feature.

Therefore, an exaptation is an abrupt change of use of a feature in a species, and this sudden event cannot be attributed to progressive selection (microevolution). It is thus identified as a case of macroevolution. Obviously, it occurs in a species, not above the species level. Therefore, macroevolution does not depend on natural selection; it depends on fortuity, on a lucky event (happen stance) of a feature that had been selected for a specific use but could be useful in another way – for another utility. That is, it could have had a posterior effect. Thus, *in our new definition,* **macroevolution is an abrupt change of use of one or more features** in taxonomic evolution not related to the progressive natural selection of these features.

In the field of genetics, macroevolution may be identified in at least one case: the gene duplication event, with a new function (effect) of the duplicate or surplus copy, called redundant stretch of DNA (Freeman and Herron 2001: 85). Because the original copy (parent copy) still produces a normal – that is, original – product, the supernumerary copy or redundant sequence is free to accumulate mutations without consequences for the phenotype (Freeman and Herron 2001: 85). The new sequence may get new effect(s) over time. Thereby it becomes a new locus (Freeman and Herron 2001: 85). This is the first mechanism scientists encountered that resulted in entirely new possibilities for gene function (Freeman and Herron 2001: 86), that Williams (1966) would have called an effect. Therefore, for us this is a case of macroevolution that is congruent with our new definition of macroevolution, but belongs to biological genetics. Gene duplication was already regarded as attributable to macroevolution, but without confidence (*e.g.* Roth *et al.* 2007: 68–69). The fact that gene duplication with change in the products of the new copy is not currently being clearly regarded as a case of macroevolution by geneticists is possibly a result of the unclear original concept of macroevolution, grounded on the fact that Hennigian phylogenetic systematics or cladistics has not yet been aimed strongly enough by other biologists except by the systematists themselves. Indeed, we apparently are the first authors to link macroevolution to cladistics, by way of the concept of exaptation, originated by Williams (1966) and named by Gould and Vrba (1982), but see McLennan (2008: 249). Symptomatic is the fact that the books of Williams (1966) and Hennig (1966) were published at the same time and one was not aware of the work of the other. The first edition of the book of Hennig (1950) was published only in German. Therefore, it is reasonable to believe that their ideas ran in parallel in the history of biology, converging only here in our book, as far as we know. There are other ways of perceiving the synchronicity of parallel events in biology. One known mechanism for gene duplication is the asymmetry in crossing-over. Other mechanisms that can create new genes or radically new functions for duplicated genes are overprinting (Freeman and Herron 2001: 88) and reverse transcriptase, where the new locus may be functional (Freeman and Herron 2001: 89). More recently, other types of gene duplication have been described, such as segmental duplication, tandem duplication, and retroposition (*e.g.* Kong *et al.* 2007) and maybe others (*e.g.* Voordeckers *et al.* 2012). Whether these changes in DNA that produce new genes may be considered by-products (the "spandrels" of Gould and Lewontin 1979) derived from the dynamics of the genome (Freeman and Herron 2001: 90) might be debated by geneticists. According to Freeman and Herron (2001: 86) the globin gene family provides an excellent example of how duplicated genes diverge in function (Freeman and Herron 2001: 86). However, our book is focused on anatomy and the passage of vertebrates from water to land, so we will not deal further with the field of genetics.

Our new definition of macroevolution is by no means meant to prevent other new, parallel definitions. It only reflects our perception that there is a connection of the concept of exaptation with abrupt evolution and therefore macroevolution, as opposed to the progressive evolution by natural selection, that is usually linked to the term microevolution.

Gould and Vrba believe that the "adaptationist programme" (Gould and Lewontin 1979) of modern evolutionary thinking has been weakened as a result of challenges from all levels, from molecules to macroevolution. At the biochemical level, we have neutral theories and suggestions that substantial quantities of DNA may not be adaptive at the level of the phenotype. Researchers in macroevolution argue that adaptations in populations lead to effects that produce the diversification patterns of a species. According to them, such patterns may result in evolutionary tendencies (Gould and Vrba 1982). On the other hand, the pluralist view of evolution (Gould and Lewontin 1979) has been increasingly supported (*e.g.* Nelson*et al.* 1984, Bittencout-dos-Santos and El-Hani 2013, La Porta *et al.* 2020, Linde-Medina 2020, Ellis *et al.* 2022, Frenkel-Pinter *et al.* 2022).

Also, macroevolution has been correlated to a series of rapid transformations followed by long periods of very slow changes, or stasis (Allen 2014: 25), the so-called punctuated equilibrium. A macroevolutionary framework has been correlated to comparisons among species encompassing millions of years of evolutionary time (Larson 2014: 89). The term macroevolution often carries connotations of a rejection of Darwinian evolutionary theory or at least of the perception that Darwinian theory is inadequate to explain major features of evolution (Larson 2014: 89). Macroevolution has also been correlated with the ecological theory of adaptive radiations (Harmon 2014: 572). Benton (2014: 585) suggests that macroevolution may be the result of pluralistic interactions, with biotic interactions shaping ecosystem and species evolution on a daily and yearly basis, and abiotic interactions acting over all time-scales (mainly centuries and millions of years). In contrast to microevolution, the term "macroevolution" is associated with large scale evolutionary events, occurring above the species level (*e.g.* Benton 2014: 579). For example, as used to explain the appearance of a new higher group, such as the vertebrates (Ridley 2004: 540). Yet, evolution only occurs during the lifetime of individuals. Evolution occurs concomitantly along generations of these same individuals, that represent living species. With the introduction of phylogenetic reasoning, monophyletic groups above the species also represent real entities, individual systems equivalent to individuals and species, in the sense that the evolutionary process or processes that occurred in the exclusive, ancestral species of all the remaining species that belong to the monophyletic group is or are the evolutionary process(es) that influenced all of them, directly or indirectly. Harmon (2014: 572) predicted that our current improved knowledge of phylogenetic relationships between taxonomic groups will shed light on the issue of macroevolution. As authors, we believe this book contributes to this perception of Harmon. Monophyletic groups include an ancestral species and all of its descendants. Thus, evolutionary processes may be understood studying the ancestral species of each monophyletic clade, by means of its apomorphies. Supraspecific entities which do not correspond to monophyletic entities become abstractions not subject to the evolutionary process. Considering a paraphyletic and mainly a polyphyletic group, different original species are implied and, therefore, different evolutionary processes are possibly involved. In this sense, it may be said that these non-monophyletic groups are not subject to an evolutionary process. Evolution proceeds at the species level, or of the ancestral species and all of its descendants, in the case of a supraspecific clade. Consequently, the term "macroevolution", as originally proposed, loses its meaning, as all evolutionary processes automatically represent "microevolutionary" events. We believe, however, that the term "macroevolution" can be saved when the original intention of the term, that of a "rupture" in the continuous and gradual evolutionary process of natural selection, is meant.

In this book, we utilize "macroevolution" as an evolutionary process in which an evolutionary novelty generates not only an adaptive advantage, but also a new, alternative possibility of interaction with the environment, parallel to that for which the novelty was originally selected. When such a possibility is performed later in the evolutionary scale, after other necessary characters have appeared for working together, macroevolutionary thinking regains validity.

This type of chance event differs from chance variations of gene alleles, which normally predominate in textbooks and research programs in this area of knowledge. The chance events we are referring to involve a feature (structure, trait), to be used for another use, different from that function for which it was initially selected. Even when the selective value is neutral at the origin of the character, this character may undertake a very important role in the life style later in the evolution of the taxon, enabling a new or more efficient mode of interaction with the environment. In the strictest sense, macroevolution is exaptation (at least if a new life style is generated), but macroevolution becomes more noticeable when there is a collective exaptation at once, generating the new life style or new mode of life.

For example, a "fish" in movement uses the head first to contact the water, and to move the water aside, in order to pass through it. This animal has an internal skeleton inside the head (the skull) to give physical support against the surrounding water. The skull is thus very important hydrodynamically. Originally, the cranium (roughly equivalent to the skull) appeared as an enlarged cartilaginous channel to accommodate the cerebrum. This is seen in the "fish" or fish-like animal called hagfish. The cranium was located just below the brain or cerebrum. At the same time, it "emitted" elongate, trabecular prolongations, to support the gills (*e.g.* Janvier 1996: 45, *fig.* 3-3 B1–12). This is the extent of the development of the skull in the hagfish, the sister group of the vertebrates. At this level of evolution, the cranium did not protect the brain laterally or dorsally. Only later in the evolution of the vertebrates did it also serve to protect the cerebrum. A hydrodynamic function apparently occurred even later, together with an enhanced swimming function. Powerful swimming literally demanded hydrodynamic efficiency of the cranium. So, natural selection may have been involved as the microevolutionary process responsible for generating a hydrodynamic and hard skull to counteract water resistance. Without the previous development of a simple cranium for other functions, this later development would not have been possible. The original function of the simple cranium may have been only to stabilize the branchial arches, that in turn support the gills. This is what we call macroevolution. There was a change in function, shape, and size. New evolutionary possibilities appeared, but, in this case, these did not evolve or develop progressively and uniformly, as predicted by "Darwinian" evolution. Without evolutionary "chance", without the casual appearance of a cartilaginous channel between the cerebrum and the elongated bars to support the gills, a cranium in the ordinary understanding of this term would not have appeared. We refer to this phenomenon as macroevolution throughout this book.

We thus use macroevolution to label a rupture in an otherwise continuous evolutionary process, because the feature gained by natural selection or by genetic drift could be used in new functions, or new interactions of characters with their environment. Exaptations thus do not depend on natural selection for the secondary use or effect.

We also use macroevolution for a situation in which several distinct characters appeared independently by natural selection, or as "spandrels" (by-product), or genetic drift, or other ways of evolution, but that jointly permit a new possibility of interaction with the environment. The independent evolution of several characters united into a newly functional set of features, or providing a new way of life (or lifestyle), permits us to say that this new set of characters was acquired fortuitously by the taxon. These features were exapted collectively, and the macroevolution in this case was possible only by the chance event of emergence of each and all necessary features, characters, or traits, thus characterizing macroevolution. Without a single one of them, macroevolution would not have occurred, because this was not a microevolutionary event, a step-by-step, a feature-by-feature, process of natural selection. That is the explanation for why some taxa have several supposed preadaptations for a given lifestyle, but never get this given, new lifestyle (or sometimes even manage to attain a completely different way of life), that indeed was attained by other taxa. We will return to this theme, mainly in Chapter 9.

Paraphrasing Jablonski (2005) on mass extinctions, we may state that with our new proposed definition on macroevolution, exaptation is an important challenge to the extrapolationist premise that all macroevolutionary processes are simply smooth extensions of microevolution, mainly when a group of features exapt collectively.

Macroevolution, as defined in this book, is related to the concept of exaptation. Further, macroevolution may be sometimes related to fortuitous sets formed by exaptations. Such exaptations may be either correlated or then not correlated among each other. Perhaps they are only able to work together as a complete set of interconnected operating units that permit the appearance of new life styles. The new mode of life may depend on the available complete set of characters (traits) workable in the new way as a whole. Such exaptations may have appeared fortuitously regarding each other. The lack of a single feature from the necessary set of needed features (traits) simply prevents the collective exaptation of the set for the new lifestyle or new way of life. When a set of features exapt together, then the chance event represents a sum of fortuities. As may be seen from our previous exposition on exaptation, the fortuitous nature of the exaptation itself is clear, because natural selection for a given environmental pressure may select different features in different taxa: the thermal body insulation in birds (feathers) is not the same as in mammals (hairs), in a simple example. Furthermore, in this example, feathers could be exapted for flight while hairs could not. The process of macroevolution thus becomes characterized by the fortuitous and independent nature of its different and independent exapted features or traits.

CHAPTER 4

Monophyly and Geological Time as Counter-arguments to the Adaptationist View of the Origin of Terrestrial Vertebrates

The evidence that adaptationist biologists use may lead them to incorrectly classify a trait (a character or a set of characters) as an adaptation for a certain life function of the animal (Gould and Lewontin 1979 *ap*. Andrews *et al.* 2002: 498). For example, genes underlying the trait could have evolved by chance, by mutation, or the trait could be the secondary product of a certain adaptation ("spandrel", in the sense of Gould and Lewontin 1979). In other words, a character is an adaptation for one function, which was subsequently exapted for the proposed function, among many other possibilities. Williams (1966: 12) says that a frequent practice is to consider any recognizable benefit that may result from the activities of an organism as an adaptation. This is an insufficient basis to propose an adaptation and has led to some serious errors. A benefit may be the result of chance, instead of having been selected ("designed" in an evolutionary sense). Williams (1966) holds the view that researchers should make an inference that adaptation has occurred only after demonstrating that all alternative hypotheses to adaptation are highly improbable as full explanations for the effect of a trait. According to Williams (1966), the demonstration that adaptation has occurred carries the burden of proof. The principle of adaptation should only be used as a last resource. It should be emphasized that authors such as Williams (1966) do not infer that adaptation does not occur, but only imply that other evolutionary processes may act as well. These may be under evaluated or underestimated, when they occasionally may represent the main evolutionary event in a certain taxon or in a particular context (*e.g.* Gould 1997). Darwin (1872 *ap*. Gould 1997) himself said that "I am convinced that natural selection is the main, but not the only, means of change".

1 Monophyly of Terrestrial Vertebrates

Triques and Christoffersen (2009) used a cladistic structure to argue that adaptive scenarios for selective forces are not compatible with the appearance of a single monophyletic group of vertebrates that colonized land. Vertebrates have an exclusive type of limbs and a unique structure representing the foot.

 | DOI:10.1163/9789004684911_005

The monophyly of both Sarcopterygii and Tetrapoda, and also of terrestrial tetrapods as well (which exclude the first tetrapods, all of which were aquatic) offers weak support for considering that microevolution has been responsible for the invasion of land. Why would a single structure have appeared by natural selection to support the body weight, and why such a unique structure as the limbs and the foot? That is, if selective pressures were the evolutionary mechanism leading to the conquest of land by vertebrates, then it would be expected that more than one kind of limb and more than one kind of foot would have been developed by natural selection, such as the case of the different kinds of wings found in the three groups of aerial vertebrates, the pterosaurs, birds and bats. The pterosaur wing is based on one very elongated finger that supports a dermic membrane. That of birds is based on elongated wrist bones and two not very elongated fingers together with feathers all through the structure of the limb. Finally, that of bats is based on four very elongated fingers that support a dermic membrane. In contrast, the limbs and feet of the most ancient tetrapods have basically the same structure as arms.

The logic of this reasoning is that if evolutionary pressures in fact exist, and if evolutionary advantages in fact explain the appearance of certain characters or life styles (terrestrial life, for example), then all organisms living in the same environment would have been selected in the same way, so that several analogous and functionally equivalent characters would have been selected independently of each other, approximately at the same geological moment, but clearly different from each other, anatomically. However, terrestrial vertebrates appear only once (that is, they form a single evolutionary lineage, rather than several independent evolutionary events). In other words, they form a single monophyletic group, instead of being polyphyletic (with different evolutionary origins), as would be expected if they had originated as a function of selective pressures for the conquest of land. Taking the above example of aerial vertebrates, they do not form a monophyletic assemblage. Pterosaurs are usually considered to be close to or the sister group of dinosaurs. Birds are a derived group (or a subgroup) of dinosaurs. And bats are mammals.

When the environment favors the adaptive selection of certain forms in order to better explore a particular life style, the final expression of those forms may be traced back to different original lineages (*e.g.* Gutarra and Rahman 2022: 69, *fig.* 1). In fact, some of these adaptations have long been used in textbooks (*e.g.* Newman 1923: 17, *fig.* 4; McFarland *et al.* 1979: 63, *fig.* 3-3; Hildebrand 1982: 17, *fig.* 1-5).

For example, hydrodynamic vertebrates (those animals with a somewhat fusiform body) are notorious for having evolved separately several times. Animals as different as ichthyosaurs, the white shark, and its relatives (lamnid sharks together with *Alopias* – thresher sharks – and *Megachasma* – the megamouth [*e.g.* Shirai 1996: 20]), dolphins and tuna fish, are used as examples of analogous evolution (in opposition to homologous evolution) of the body form (*fig.* 4-1). In fact, the origin of hydrodynamism in Teleostei (Actinopterygii) occurred so many times, independently, that perhaps only a consortium of specialists would be able to establish the approximate number of parallel, convergent, events. As examples, we may consider the following teleost fishes that independently gained an approximately fusiform body shape along the evolution of their lineages: Scombroidei (tuna fish and their relatives, *fig.* 4-1 A), Exocetidae (flying fish, *fig.* 1-3 illustrates an Atherinomorpha representative), Mugilidae (mullets and paratis, *fig.* 4-1 E), Salmonidae (salmons and trouts), Esocidae (pikes), Megalopidae (tarpons, *fig.* 4-1 D), among many other examples. More basal in actinopterygians, and also hydrodynamic, are the gars or alligator-fish (Lepisosteiformes), *fig.* 4-1 G. Further, we may indicate separate origins for the fusiform body shape in several subgroups of chondrichthyans, such as Lamnidae and related forms (already mentioned above, *fig.* 4-1 B), Carcharhinidae (but possibly independently within the different genera), *Rhincodon typus* (the whale-shark) and †Cladoselachidae (*e.g.* Shirai 1996: 20). The sarcopterygian †*Eusthenopteron* (torpedo-shaped sarcorpterygian, *fig.* 5-2), belonging to †"Osteolepiformes" (*fig.* 5-1) and the vertebrates without jaws of the group †Anaspida are also approximately fusiform. Even the penguins (Aves, Sphenisciformes) are relatively fusiform birds, if we refer to their shape when swimming. The group of mammals (Mammalia) produced highly hydrodynamical forms, such as cetaceans, which include whales and dolphins (*fig.* 4-1 F). The †ichthyosaurs, fossil marine "reptiles" well known to paleontologists (*fig.* 4-1 C), are not clearly positioned phylogenetically, but may be close to Lepidosauria or Crocodilia + Aves (*fig.* 1-4). Both cetaceans and †ichthyosaurs returned to water, originating from totally terrestrial vertebrates, being subgroups of Amniota (*fig.* 1-4). The occurrence of relatively fusiform groups within all the above cited taxa indicates that microevolution may have been responsible for the evolution of each group. These groups may be considered evolutionary lineages that were under selective forces to better explore their environment. Selection for speed in open water produced this shape independently (at least on the basis of currently known phylogenies). Thus, if fusiform vertebrates were to be united, they would form a polyphyletic assemblage. This is so clear that such a taxonomic grouping has never been proposed.

Another example of multiple conquests of an environment is known for a long time: the return of tetrapods to water – presumably an advantageous life style under many circumstances – from evolutionary lineages that

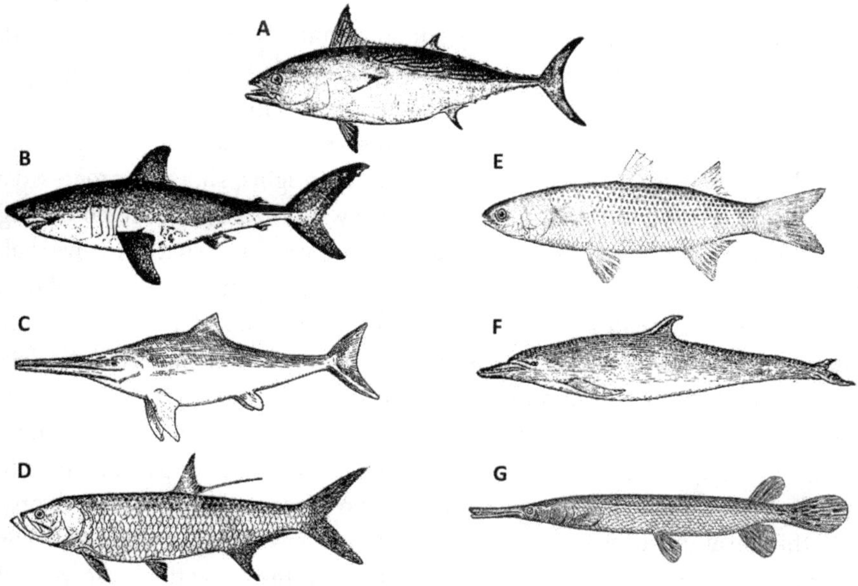

FIGURE 4-1 Illustration of hydrodynamic vertebrates, of fusiform shape or very similar to fusiform shape. A – Tuna fish (Actinopterygii, Teleostei, Scombroidei, Scombridae). B – Shark of the same family as the White Shark (Chondrichthyes, Lamniformes, Lamnidae). C – †Ichthyosaur (Sarcopterygii, Tetrapoda, Amniota). D – Tarpon (Actinopterygii, Teleostei, Elopomorpha, Megalopidae). E – Mullet (Actinopterygii, Teleostei, Mugilomorpha or Percomorpha, Mugilidae). F – Dolphin (Sarcopterygii, Tetrapoda, Amniota, Synapsida, Mammalia, Cetacea). G – Alligator-fish or gar (Actinopterygii, Lepisosteiformes). A – *Thunnus* sp., from Jordan & Evermann (1902). B – *Lamna cornubica*, from Goodrich (1909). C – from Newman (1923). D – *Megalops atlanticus*, from Newman (1923). E – *Mugil cephalus*, from Goodrich (1909). F – *Delphinus delphis*, from Newman (1923). G – *Lepisosteus platystomus*, from Newman (1923)

were previously well established on land. This evolutionary tendency includes many unrelated tetrapod lineages (*e.g.* Pough *et al.* 2005; Pough and Janis 2019: 344–349): †ichthyosaurs (†Ichthyosauria) (*fig.* 4-1 C), †thalattosaurs (†Thalattosauria), †placodonts + †pachypleurosaurs + †nothosaurs +† plesiosaurs (†Placodontia, [†Pachypleurosauria + †Nothosauria] and †Plesiosauria respectively, in a currently accepted phylogenetic hypothesis) (*e.g.* Benton 2015: 150–154, box 6.1), the lizards-related †mosasaurs (†Mosasauria, *e.g.* Carroll 1988: 232–234) and †*Hovasaurus*, another aquatic diapsid (Currie 1981), the tuatara-related †pleurosaurs (†Pleurosauridae, *e.g.* Carroll 1988: 227), some aquatic turtles and tortoises (Testudinata or Chelonia, *e.g.* Benton 2015: 241), cetaceans, including whales and dolphins (Cetacea) (*fig.* 4-1 F), sirenians, including the manatees and their like (Sirenia), etc. †Mesosaurs (†Mesosauridae) are either the first-known fully aquatic amniotes (*e.g.* Benton 2015: 126), the sister group of †Ichthyosaurs (Maisch 2010: 191, *fig.* 1), or yet other hypotheses (*e.g.* Maisch 2010: 192: *fig.* 2). Also, †metriorhynchids were marine crocodylomorphs (Pough and Janis 2019: 349). In many of these lineages, the fossil record has been traced back to their terrestrial origins, showing progressive changes for the aquatic environment (*e.g.* Thewissen 2014). We are not listing here the anamniote returns to water, as they are yet somewhat controversial. The number of returns of amniotes to water is even larger than we listed above, and yet this is enough to present the reader to a group of phylogenetically unrelated animals. This situation is exactly what would be expected if selective pressures for gaining another way of life were involved. The case of the terrestrial origin in vertebrates is precisely the opposite situation. If a taxonomic group including all these vertebrate forms that have returned to an aquatic life style had been joined taxonomically, it would represent a polyphyletic taxon.

A third example of selective pressures acting in favor of adaptation to a new environment seems to be that of depressiform fish (involving the dorso-ventral compression of the body shape). Squaliform sharks apparently become progressively more flattened, starting with the loss of the anal fin (*e.g.* clade 19 of Shirai 1996: 20, character 84), with a subsequent flattening of the caudal peduncle, and finally with the appearance of benthic habits (even though some aggressive predators living in the water column are known in this group, such as *Isistius brasiliensis* – the cigar shark). This apparent evolutionary tendency for the body to become dorso-ventrally flattened and for the fish to acquire a benthic mode of life is also seen in Hypnosqualea (De Carvalho 1996: 38, *fig.* 5), that includes the angel sharks, sawsharks, sawfishes and rays, stingrays, skates, manta rays, etc. This evolutionary tendency is supposedly followed by a change in habitat in some rays, including the manta rays, which live just below the surface of the sea. Many independent origins for this body shape include the †rhenanid †placoderms, several families and genera of siluriform teleosts (Chacidae, Aspredinidae, *Loricaria, Rineloricaria, Lophiosilurus, Hemiodontichthys, Sturisoma*, etc.) becoming independently depressiform (*e.g.* de Pinna 1998: 289, *fig.* 1), the group of the angler fish (Actinopterygii, Teleostei), Ogcocephalidae (*fig.* 3-1), Lophiidae, Batrachoididae, among many others. Thus, the contrast of the unique origin of terrestrial vertebrates with the diverse origins of depressiform vertebrates becomes evident. Molecular phylogenies of chondrichthyans that show skates, rays, manta rays etc. as a non-related group to the depressiform angel sharks, sawsharks and sawfishes (*e.g.* Human *et al.* 2006) "creates" at least one more taxonomic depressiform group or lineage, and therefore only strengthens our proposal. That is, the whole group formed by depressiform vertebrates would be a polyphyletic assemblage, if it were ever erected as a formal taxonomic unit.

Another counter-argument can be found in Wiens *et al.* (2006: 132, *fig.* 3), on the evolution of limb-reduced "ecomorphs" found in a phylogenetic "supertree" for squamate "reptiles" (all quotation marks are ours). Reconstructed using Pagel's maximum-likelihood method, at least 14 independent emergences of this character were found. Indeed, the authors state that the snakelike body form has evolved about 25 times (Wiens *et al.* 2006: 123). Most of these groups belong to one of two "ecomorphs" (quotation marks from the original authors), either short-tailed burrowers, or long-tailed surface dwellers.

Thus, a foot appeared only once in vertebrates, and the occupation of dry land occurred as a single evolutionary event in vertebrates having feet, and, indeed, in vertebrates as a whole. These unique origins of the foot and of the conquest of land contrast with the numerous acquisitions of hydrodynamic bodies for better movement in the water column, with the diverse returns of terrestrial vertebrates to aquatic environments, and with many evolutionary experiments in occupying flat environments (such as shallow waters, sea bottoms, oceanic surfaces, etc.) by several groups of vertebrates previously living in the water column in previous evolutionary stages, as well as by the numerous cases of the limb-reductions in vertebrates. These contrasting examples corroborate the previous proposal of Triques and Christoffersen (2009) that a process of adaptation for life on land seems improbable for the vertebrates. Thus, a macroevolutionary process is apparently involved, rather than a microevolutionary event.

Consider the cladogram of gnathostomes in *fig.* 1-2. Terrestrial vertebrates are a subgroup of Sarcopterygii (see *fig.* 1-4). If selective pressures were the mechanism directing the conquest of land by vertebrates, all gnathostomes through Silurian and Devonian times would have to be subject to these selective pressures, because all of them were aquatic, occurring in salt and fresh water. Besides Sarcopterygii, there were four other large taxonomic groups subject to the same environmental conditions/pressures: †Placodermi, Chondrichthyes, †Acanthodii, and Actinopterygii. Why did only Sarcopterygii develop terrestrial forms, and for that matter, only a single terrestrial group of vertebrates? In other words, if selective pressures were the leading forces for terrestriality in vertebrates, the terrestrial vertebrates would form a polyphyletic group, originated in at least some other taxonomic group lying outside the sarcopterygians.

2 The Geologic Time Gap

Geological time may also be invoked in favor of the idea of macroevolution, at least at the origin of terrestriality in vertebrates. Marine sediments found in

Poland, dating from the Middle Devonian (Eifelian Stage), are approximately 18 Myr older than the oldest known fossil tetrapods, and appeared at least 10 Myr before the fish-like sarcopterygians closely related to the tetrapods, which are placed in the †"Panderichthyidae". The quotation marks indicate that its genera lie progressively closer to Tetrapoda in current phylogenies, instead of having synapomorphies that justify their monophyly. The †panderichthyids include †*Panderichthys*, †*Elpistostege* and †*Tiktaalik*, being simply represented as †"Panderichthyidae" in *fig.* 5-1. The Polish fossil trails referred to above represent the oldest known fossil vestiges of vertebrates with feet. They include fossil trails of footprints, apparently belonging to tetrapods. These Polish records thus indicate the oldest known tetrapod vestiges. They force us to radically reevaluate the geological, ecological, and environmental context of the fish-tetrapod transition. They also make us wonder to what degree the fossil record may be considered complete (Niedzwiedzki *et al.* 2010: 43). These trails are 395 Myr old. In fact, they further corroborate the ideas of Triques and Christoffersen (2009), that microevolution fails to support the origin of terrestriality in the Tetrapoda. Niedzwiedzki *et al.* (2010) have found fossil evidence that extends our present knowledge on the occurrence of fossil tetrapods to 18 Myr before previously published time of origin. There has thus been much more geological time available to refine tetrapods for life on land. A longer time for natural selection to proceed, however, makes microevolution less likely to be the causal factor responsible for the conquest of land. In fact, if selective pressures were involved, why would such an extended time be necessary for the implementation of microevolution? For instance, according to the evaluation of Thevissen (2014), all the transformation of Cetacea from terrestrial animal to fully aquatic forms took only eight Myr. These new fossil discoveries of Polish tetrapod trails include well-preserved trails of "hands" and feet. Some of these footprints are very congruent with what is already known for †*Acanthostega* (the oldest known vertebrate with an almost complete skeleton, *fig.* 5-16), and particularly for †*Ichthyostega* (Niedzwiedzki *et al.* 2010: 46), indicating the possibility of a common origin of these lineages. These possibilities again corroborate the hypotheses of Triques and Christoffersen (2009) that microevolution was not the factor involved in the unique origin of the terrestriality in the tetrapods. Before the discovery of the Polish fossils it would be possible to consider a period around 25 Myr from the origin of Tetrapoda until the occupation of land (Benton 2005: 85). With the Polish fossil, this period may reach possibly 35 million years (McGhee 2013: 32, tab. 2.2) or even 41 million years (Benton 2005: 27, box 2.2), depending on where we establish the beginning of the Carboniferous – usually identified as the possible date of origin of terrestriality in vertebrates. These extended spans of geological time are not congruent with explanations of the origin of terrestriality based on microevolution and are inconsistent with the proposed evolutionary scenarios for the transition from water to land, particularly when we now know that these first vertebrates were already capable of producing footprints. In other words, feet and toes were already directed downwards so that the whole foot could be forced against the soil and was capable of producing impressions in the wet mud. Thus, feet were basically ready for walking and yet the conquest of land still required another 35–41 million years. This time interval seems incongruent with the idea of an evolutionary advantage that follows from natural selection, from adaptationism, or of the idea of microevolution, opening the door for a macroevolutionary view of the origin of terrestrial vertebrates. These points only strengthen the hypothesis of Triques and Christoffersen (2009) that alternative macroevolutionary mechanisms must be sought. The presumed walking ability, with a diagonal pattern of steps in Poland, seems to be in accord with the peculiar alternate movements of *Latimeria* (Coelacanthiformes or Actinistia), as observed from a submersible vehicle (Janvier 1996: 74; Friecke *et al.* 1987; Uken and Green 2006: 475). This pattern is also observed in the movement of the fins in living lungfish (Dipnoi) (Janvier 1996: 74–75). Although *Latimeria* and lungfish may also move their paired fins as terrestrial vertebrates move their legs, these fish-like sarcopterygians do not walk on the bottom of water bodies. It is possible that the fossil species that left footprints in Poland only "walked" sporadically, not as a usual way of life. This trait could be neutral regarding natural selection. We are going to return to this feature as character 2 of the cladogram in *fig.* 5-1, in Chapter 5. We are going to present it in more detail and discuss it in terms either of adaptation for a function, or other possible interpretations. Thus, the pattern of movement of the paired locomotor appendages (that in these animals are in the shape of fins), to produce "walking" in completely aquatic vertebrates, and known at least since the end of the Silurian, is similar to the pattern of walking of present-day terrestrial vertebrates.

Even if the Polish trails are shown in the future to be made by one currently unknown tetrapod taxon, or yet a taxon with feet of completely different phylogenetic origin regarding Tetrapoda, our argument is only strengthened because a new evolutionary lineage would be shown to have had the opportunity of being tinkered with natural selection and to conquer land, which never occurred in fact.

After the publication of Triques and Christoffersen (2009), which argued in favor of macroevolution and of exaptation in the origin of terrestrial vertebrates, larger works began to be published, containing opinions that admit that effectively terrestrial vertebrates must have involved several exaptations (*e.g.* Laurin 2010: 35, 88, 139–140, 171, Schoch 2014: 10). The terrestrial invasion is considered by McGhee (2013) to have been a very improbable event, very close to not having occurred, as may also be deduced from the approach of Triques and Christoffersen (2009). Despite authors such as these, which promote more liberal views diverging from adaptational constraints, the absolute preponderance of exaptations, as we are going to show in Chapter 5, may surprise many authors, given their indiscriminate preference for adaptive scenarios provided as explanations for all and any cases of evolution within the vertebrates.

Taking the monophyly of the terrestrial vertebrates into account, the number of terminals that could have led to land vertebrates can be discussed in at least two differing directions. One has to do with the number of different evolutionary lineages, each one with its own basic plans ("ground plans"). In this respect, some features occurring in, and sometimes defining the lineages, may dictate possibilities and impossibilities, even if relative. For instance, the heavy skeletons present in most †placoderms (it does not matter if this group is considered a monophyletic or paraphyletic group) would constitute a negative handicap for terrestriality. On the other hand, as will be discussed in detail in Chapter 5, a single skeletal element proximally in the paired appendages permits a rotational connection of each leg with each of the body girdles, allowing the necessary step-like movement of the four legs for walking. In this evolutionary context, the number of different evolutionary lines is or may be a differential in the global analysis of gnathostome vertebrates.

The other direction of argumentation is simply the number of species of gnathostome vertebrates occurring in a horizontal cut of geological time. In this regard, the number of actinopterygian "fishes" in the last million years is striking and demands some considerations.

3 The Extreme Diversity of the Actinopterygii

According to Janvier (1996: 183), Actinopterygii present one of the largest and most diverse extant vertebrate groups, but they did not gain this great diversity before Triassic or Jurassic times, and they first became relatively abundant by Early Carboniferous times. The earliest remains regarded to represent actinopterygian animals are isolated scales, teeth and dermal bone fragments from the Later Silurian of the Baltic area (Janvier 1996: 183). Janvier (1996: 184) considers that the first large actinopterygian diversification ("radiation", as he refers to) took place by the end of the Frasnian. Frasnian is a "stage" in the Devonian period from 382 to 372 Myr, said to be upper Devonian, but not attaining the limit with the Carboniferous period. The Frasnian is a space of time at the very beginning of the Late Devonian, when Tetrapoda seem to have appeared. The end of the Frasnian is a time space from about 375 to 373 Myr, when Tetrapoda are supposed to have just emerged (*e.g.* Benton 2015: 89). Before this time, actinopterygians are rare and rather homogeneous in aspect (Janvier 1996: 184). If Janvier's evaluation (1996) is correct, the first great diversification of actinopterygians occurred almost at the same time in which the tetrapod sarcopterygians had just started to diversify. Therefore, a significant splitting of Actinopterygii occurred still in the Devonian that were under the same "evolutionary pressures" – whatever this may mean for adapationists – as the emergence of tetrapods from aquatic environments. If evolution is only meant to occur according to adaptation to selective forces, then there is no other way to be other than aquatic vertebrates living at the same time and under the very same "selective forces". It should expect that all vertebrate forms of the time were under the same evolutionary forces and should respond in the same way or direction. However, Actinoterygii did not lead to true terrestrial forms. Exactly when actinopterygians were becoming more heterogeneous in aspect ("radiating", so to speak), under the same Devonian conditions that supposedly led sarcopterygians "toward" land (Late Devonian), the same did not occur with actinopterygians, that never originated a terrestrial group of vertebrates in the same sense of the Neotetrapoda (the current formal name of the crown group Tetrapoda). Not even the argument that terrestrial vertebrates were already on land, preventing by competition a new wave of terrestrial vertebrates of another origin could be invoked, as the sarcopterygians had still not invaded land by this time. It is important to remember that the Devonian tetrapods were all aquatic forms.

In the evaluation of Janvier (1996: 188), Silurian and Devonian actinopterygians are few [in species numbers, it seems for us from the statement of Janvier], and most Devonian forms are quite similar in general aspect. At this point, it seems to us that Janvier is referring to Lower Devonian because in the next paragraph he states that by Middle Devonian times, particularly of Givetian age (about 387 to 383 Myr as in Benton 2015: 89) – which is the end of the Middle Devonian, and right before Frasnian – the diversity of actinopterygians increased, with some forms reaching 70 cm in length. He continues commenting on Devonian actinopterygians when he states that in the Famennian (the last epoch of the Devonian period), advanced forms appeared, some of them possessing deep bodies and/or crushing teeth. Carboniferous and Permian actinopterygians display a wider range of diversity, with eel-shaped or deep-bodied forms (Janvier 1996: 188). Therefore, we may summarize on actinopterygians from Janvier (1996). They appeared in the Silurian, diversified only slightly in species groups, having a "small" body size (about 10 cm, as stated for a more advanced genus, †*Mimia* [Janvier 1996: 184]). In the Lower Devonian and in the Middle Devonian, Actinopterygii diversified largely into different subgroups. Morphological diversification of the body as well as size varied along this period, and also through the Upper Devonian. In the Carboniferous and Permian periods, the diversification continued. Exactly at the same time of the strong actinopterygian diversification, the tetrapod diversification occurred. The first group (actinopterygians) did not lead to true or complete terrestrial forms but the second group (tetrapod sarcopterygians) did.

In order to make clear our argument against adaptationism of vertebrates at the origin of terrestriality, it is necessary to highlight that several basal actinopterygians up to the basal level of Teleostei (see *fig.* 1-3) have lungs or a respiratory gas bladder (in the cases where a gas bladder is said to occur, in basal Teleostei). Romer (1966: 51: *fig.* 73 A) shows diagrammatic lateral views and cross-sections of lungs in tetrapods, African and South American lungfish (Dipnoi) and the genus *Polypterus* (Cladistia, a basal living representative of Actinistia – *fig.* 1-3). In all these vertebrates, the lungs emerge ventrally from the gut, are paired and placed one on each side of the gut. In the Australian lungfish, the paired lungs emerge from the ventral region of the gut, on the right side, but is positioned above the gut. In "most actinopterygians", there is a "lung-air bladder" [quotation marks are ours] (as Romer 1966: 52, *fig.* 73 called the structure), usually named by other researchers a gas bladder, an air bladder, or a swimbladder). *Amia calva*, the bowfin, is the living representative of Amiiformes, and is here accepted as the sister group of Teleostei (*fig.* 1-3). It is able to breathe air and possesses a lung that arises as a dorsal diverticulum of the pharynx, in a similar fashion to the lung of the gar, *Lepisosteus*, the living representative of Lepisosteiformes (see *fig.* 1-3), according to Lauder and Liem (1983: 114). From these statements of Lauder and Liem (1983) about Amiidae, Lepisosteiformes, and that statement of Romer (1966) about Cladistia, it seems clear that basal actinopterygian groups had lungs, including several or perhaps all fossil forms, even if such lungs are not fossilized: the most parsimonious – that is, most congruent – hypothesis indicates this possibility. *Arapaima gigas* is one of the World's largest species of the so-scaled freshwater "fish", growing at least to 2,5 m in length (Nelson 1994: 93). It is a part of the Osteoglossomorpha, a basal group of Teleostei (*fig.* 1-3) and is known for breathing air, using a modified swimbladder as a lung (Lauder and Liem 1983: 125). Another representative of Osteoglossomorpha is *Pantodon*, which is a highly specialized member of this group, a small-bodied, surface-dwelling "fish", with an upturned mouth (as the mouth of *Arapaima*). It preys on small insects caught on the surface film and it has been suggested that it breathes air (Lauder and Liem

1983: 125). That is, representatives of Osteoglossomorpha, the most basal teleostean groups, according to Lauder and Liem (1983), have a "lung air-bladder", as Romer (1966) named the feature. Janvier (1996: 59) states that the homology of the lungs or air bladder in the sarcopterygians and actinopterygians, and even within actinopterygians, is still debated, because these organs arise from the gut differently and at different places. At this point of our argumentation this does not constitute a problem, because we are going to use the fact that all these forms present a lung or lung-like, respiratory bladder in their bodies. Our point is exactly that when actinoptergians had their first large diversification splits (large in number and in morphological differentiation) – from the end of the Middle Devonian (the Frasnian epoch), and beyond, through the end of the Devonian, Carboniferous and Permian – they had lungs just as their contemporary sarcopterygian relatives, including the basal Tetrapoda. Even so, actinopterygians did not lead evolutionarily to any group of terrestrial forms similar to the crown group Tetrapoda. There were no terrestrial vertebrates yet (up to the beginning of the Carboniferous period) that could somehow prevent by competition the emergence of a parallel terrestrial vertebrate group originating from actinopterygians. Therefore, this is another argument against adaptationist (selective) explanations for the emergence of terrestriality in the vertebrates. At the same time, it constitutes an argument favoring exaptations in the conquest of land by vertebrates.

Parallel to the occurrence of lungs in basal actinopterygians and basal teleosts, an elongate and developed skeletal support for the pectoral fin was present in Cladistia (*figs.* 1-3 and 5-13), which are the bichirs (*Polypterus*) and reedfishes (*Erpetoichthys*, = *Calamoichthys*). This has been the subject of extensive debate in the literature of the last hundred years (Lauder and Liem 1983: 108). Probably no other group of living fishes has been placed, at one time or another, in so many widely differing major groups (Lauder and Liem 1983: 108). They were included in the Sarcopterygii primarily on the basis of the structure of the pectoral fin in *Polypterus* (Lauder and Liem 1983: 108), which consists of a proximal lobe-like extension from the body and a fan-shaped distal array of fin rays (*fig.* 5-13), according to Lauder and Liem (1983: 108). The group was already put in a large taxon of its own (Brachiopterygii) and more recently polypterids have been considered to represent highly specialized survivors of basal actinopterygian fishes (Lauder and Liem 1983: 108–109). Adult *Polypterus* have a pair of highly vascularized lungs that arise ventrally from a diverticulum in the floor of the pharynx (Lauder and Liem 1983: 110). The bichir still has a kind of lobate pelvic fins, based on four parallel large radials and a strong pelvic girdle (Goodrich 1930: 191, *fig.* 205). They have more elongate pelvic bones than (but not as strong as) in †*Eusthenopteron*, a hydrodynamic sarcopterygian "fish" closer to tetrapods (*fig.* 5-12). Although *Polypterus,* and maybe other cladistian actinopterygians, have developed pectoral and pelvic locomotor appendages, together with highly vascularized lungs, living until today through hundreds of million years, no terrestrial vertebrate group emerged from cladistian actinopterygians. This is one more evidence against adaptationism at the origin of terrestriality in vertebrates. To complement the perception that pectoral fins had developed skeletal support in basal actinopterygians, the more derived (in the phylogeny of Janvier 1996: 190, *fig.* 4-70) †*Moythomasia* had several strong radials in this fin (*e.g.* Moy-Thomas and Miles 1971: 88, *fig.* 5-5). The fin consisted of a row of radials articulating with the scapulocoracoid in a more or less fan-like manner. This is similar to the descriptions in the paraphyletic family †"Palaeoniscidae", that also included forms like †*Rhadinichthys*, †*Gonatodus*, †*Nematoptychius*, †*Phanerosteon,* among many others (Moy-Thomas and Miles 1971: 91–92). Clearly, this fan-shaped skeletal support of "†palaeoniscid" pectoral fins is not elongated as that described and figured for *Polypterus.* In any case, this structure provides an inner strong support for these fins, that could be tinkered by selective forces toward land in many different basal lineages of actinopterygians, a transition that in fact never occurred.

Summarizing, lungs are present in basal actinopterygians to the most (or one of the most) basal teleost groups (Osteoglossomorpha). Together with a strong pectoral fin, it provides support for the fin rays in several independent basal actinopterygians. This provides a number of opportunities for selective pressures to promote the exit from water, or to conquest land, since the end of Middle Devonian toward the beginning of the Mesozoic era. Yet no terrestrial groups of vertebrates are known to belong to this taxonomic and evolutionary group. We consider this as an argument against adaptationism in the origin of terrestriality in vertebrates.

Another aspect of Actinopterygii that should be explored here is their taxonomic diversification. In this regard, it is possible to divide this theme into at least two different lines of argumentation. One is the number of phyletic lines of Actinopterygii, and the other is the number of species.

Our *fig.* 1-3 is a simplification of the results of Lauder and Liem (1983), excluding fossil taxa and some recent groups. In our mentioned figure it is possible to see Cladistia (that includes *Polypterus*), then Chondrostei (that includes sturgeons and paddlefishes), Lepisosteiformes (gars) and Amiidae (the bowfin) as progressively closer to living relatives of the Teleostei. Therefore, there are five living lineages, including Teleostei, in this figure. Lauder and Liem (1983) coalesce, use, and summarize cladograms and phylogenetic analyses previously published by other authors before them. Starting at this point, we can find in their cladogram on basal Actinopterygii (taking

Teleostei as one of the terminals) 10 lineages, including Teleostei (Lauder and Liem 1983: 103, *fig.* 6). Some lineages are represented only by one fossil genus: †*Cheirolepis*, †*Moythomasia*, and †*Pteronisculus*. In this phylogeny, Lepisosteiformes are called Ginglymodi and Amiidae are included in the Halecomorphi. The mentioned phylogeny of Lauder and Liem (1983) shows taxa progressively closer to Teleostei. This means that although Actinopterygii are a monophyletic group – that therefore shares the DNA of a single species –, each group is split from the remaining ones progressively. From this split, each resulting actinopterygian subgroup constitutes an evolutionary lineage that has nothing else to do with the remaining, or previous, relatively more basal, actinopterygian subgroups. The remaining subgroup finally gave origin to Teleostei. These 10 evolutionary lineages evolve independently from each other and were under selective forces to leave water and/or to conquer land, if such forces indeed existed. None of these evolutionary lineages conquered land, as we defined herein (walked, fed, and breathed on land, totally independent of water).

Janvier (1996: 190, 192, 193, *figs.* 4-70–4-72) includes several more recent publications not appearing in Lauder and Liem (1983), and is much more detailed than Lauder and Liem (1983). He also uses fossil genera, but recognizes groups of genera represented by one of them. For example, in his *fig.* 4-70, regarding interrelationships of fossil and recent actinopterygians, the more basal actinopterygian group is represented by the '†*Cheirolepis* group' (quotation marks of the author). Therefore, this is clearly thought as an evolutionary group – thus an evolutionary line – of more than one fossil genus by Janvier (1996). Including Cladistia and a "crown group Actinopteri" (that includes Teleostei), this first phylogeny of Janvier recognizes 14 evolutionary lineages, the last one (the "crown group Actinopteri") included. His next phylogeny is of the "crown group Actinopteri", that includes Chondrostei (see our *fig.* 1-3) and a "crown group Neopterygii", with eight more evolutionary lineages. In this phylogeny of the crown group Actinopteri, 21 lineages are encountered. Finally, in the phylogeny of Janvier (1996, *fig.* 4-72), we have the crown group Neopterygii, including Lepisosteiformes, Amiidae and Teleostei, with five more terminals, totalizing 25 lineages. All of them are supposedly under selective pressures for leaving water or gaining ground and none of them left terrestrial descendents capable of walking, foraging and breathing air independently of water, as in the crown group Tetrapoda.

We consider the fact that the so many independent evolutionary lineages of Actinopterygii never left a truly terrestrial vertebrate group (living or fossil) to be an argument against adaptationism for the occupation of ground habitats by the vertebrates.

Long (2011) presents a more recent overall view on the evolution of the Actinopterygii. He evaluates that the oldest actinopterygians appeared in the Late Silurian, some 420 Myr ago. Known only from scales and fragmentary bones and teeth, the first complete fossils of the group appeared some 25 million years later, belonging to *Cheirolepis*. This genus had fleshy pectoral fin lobes (as we already mentioned above), and is from the Middle to Late Devonian times. In the Late Paleozoic, actinopterygians underwent a great burst of diversification, with some 50 or more families appearing in the Carboniferous and Permian periods (Long 2011: 138), comprising hundreds of recorded species (Long 2011: 148). In order for the reader to have a better idea of what these 50 or so families mean in our argumentation, a good look at our *fig.* 1-3 is necessary. All of these 50 families lie below the level of Teleostei and include Cladistia, Chondrostei, Lepisosteiformes and Amiidae. That is, most of these 50 families are very basal in Actinopterygii phylogeny. This mean that even initially in the phylogeny of Actinopterygii – by the time of the origin of Tetrapoda as aquatic animals – actinopterygians were already diversifying, originating several families (living and fossils) that represent different evolutionary lineages exactly in the geologic period (Devonian) said to have the exact conditions for the rise of land vertebrates (see Chapter 2 for adaptive scenarios) and mainly after it (Carboniferous and Permian periods). The first terrestrial vertebrates known from fossil records are of Lower Carboniferous age. At this time Actinopterygii were under strong diversification, supposedly being under the selective pressures for leaving water or entering land. We consider that this large number of families, that are evolutionary lineages by themselves, emerging since Middle and Late Devonian, and mainly through the Carboniferous and Permian periods, corroborates our view against adaptationism. The large number of actinopterygian families that Long estimates to have existed by the Late Paleozoic would have been able to leave some truly terrestrial vertebrates. In the beginning of the Carboniferous, tetrapod sarcopterygians were only initially gaining ground, leaving ecologic possibilities for other land intruders. This simplified presentation of Long makes it easier for us to see how many possibilities there were for selective pressures to tinker aquatic vertebrate lineages independently from tetrapods and from other sarcopterygians. Yet only a single group gained ground: the crown group Tetrapoda, named Neotetrapoda. Had there been a selective process, it would be expected that other lineages not belonging to Sarcopterygii would have reached land, but this did not occur.

According to Friedman and Giles (2016), the actinopterygians (or ray-finned fishes, as the group is frequently called, regardless of cladistic precision) may reach to 30,000 living species considered taxonomically valid. This number is about half of all living vertebrate species (Friedman 2015, Henderson *et al.*

2022). Although actinopterygians are currently "dominant" in aquatic environments, during the Devonian period they were neither as abundant nor as diverse as today (Friedman 2015, Henderson *et al.* 2022). Actinopterygii seem to have been poor in species numbers at the beginning of their evolutionary history (Friedman 2015). Therefore, it is possible to conclude that this group of vertebrates increased in its number of species through time. Many genera appear progressively in the fossil record through the Devonian and Carboniferous periods and several taxonomic groups such as families or larger groups became recognizable in the fossil record of the Permian and Triassic periods, including the Teleostei in the Late Triassic (Friedman 2015). They not only survived the largest known extinction in geologic time (the Permo-Triassic extinction), but increased in species, as well as above-species, taxonomic numbers. Actinopterygians became relatively diverse after the end of the Devonian, while †placoderms became extinct. In a striking contrast, the fish-like sarcopterygians never recovered from Devonian-times (Friedman 2015: 222). For Friedman (2015: 222), it is clear that actinopterygians were the dominant "fish" group (in pre-cladistic terms) in freshwater settings by the end of the Permian. From this point of view, the number of actinopterygian species from which a new, non-sarcopterygian terrestrial taxonomic group of vertebrates could emerge, increased. Thus, Actinopterygii become the dominant aquatic vertebrate group. The group increased, especially through and after the Permo-Triassic extinction. This extinction severely affected the already established terrestrial vertebrate fauna, supposedly permitting actinopterygians to secondarily colonize dry land, and yet this never occurred. The increasing number of species and new evolutionary lineages of actinopterygians, mainly after the most devastating extinction event in Earth history, that reduced the number of species to about half of the number of vertebrate species, did not cause a secondary colonization of land by the so-called "ray-finned fishes". This is a complementary argument against the adaptationist viewpoint of the conquest of land by vertebrates. Based on divergence-time analysis of the recombination activating gene 1 (RAG 1), widely sequenced across major vertebrate groups and downloaded from GenBank for 217 gnathostome species, Alfaro *et al.* (2009) found that many gnathostome clades have relatively recent origins, with the six most species rich stem lineages originating in the last 300 million years (Myr) and the top two, modern birds (9,191 species), and percomorph fishes (15,493 species), appearing within the last 175 Myr. *Fig.* 1 of Alfaro *et al.* (2009) shows that the crown group Actinopterygii initiated their diversification roughly between 290 and 280 Myr. Teleostei started diversification bursts roughly about 200 to 190 Myr (Alfaro *et al.* 2009). Once the great Permo-Triassic extinction occurred about 250 Myr, hitting terrestrial tetrapods

30 to 40 Myr after a burst of actinopterygian diversification, it becomes difficult to explain the non-invasion of actinopterygians through land by means of pre-existing ground tetrapods, because of the strong extinction process they faced. Tetrapoda faced the strongest known extinction that hit the whole biota (animals and plants). Therefore, the increasing and diversifying actinopterygians crossed the largest known extintions (Permo-Triassic and Createceous-Paleogene), diversifying into large taxonomic units, and yet did not conquer land. Had selective pressures along geologic time been responsible for the conquest of land, it would be expected that actinopterygians would have produced some land-evolved lineages, but this did not occur.

4 The Phyletic Nature of †Placoderms and †Acanthodians

For Nelson (2006: 36), there is now strong evidence that †placoderms are monophyletic, and five features are given in Goujet and Young (2004) supporting this conclusion. The hypothesis that †placoderms are the sister group to all higher gnathostomes was followed by Nelson (2006: 36), based on Goujet and Young (2004), who pointed out that the hypothesis had already been suggested by B. Schaeffer in 1975. Nelson (2006) does not present numbers of species for the †Placodermi, neither for their subgroups. However, he follows Goujet and Young (2004), with slight modifications, in recognizing six orders, which can be considered the main evolutionary lineages: †Acanthothoraciformes, †Rhenaniformes, †Antiarchiformes (with two suborders), †Petalichthyiformes, †Ptyctodontiformes, and †Arthrodiriformes (four suborders, including †Phyllolepida, which is considered independent of arthrodires, by previous authors). Nelson (1976) already presented six orders for the †Placodermi, but arranged slightly differently: †Rhenaniformes, †Antiarchiformes, †Petalichthyiformes, †Ptyctodontiformes, †Phyllolepiformes, and †Arthrodiriformes (with two suborders). The number of evolutionary lineages remains the same.

The apomorphies supported by Goujet and Young (2004) for †Placodermi are: (1) an exoskeletal shoulder-girdle forming a rigid ring encircling the trunk, and articulating with the skull through a paired dermal neckjoint. We will consider this feature below; (2) a distinctive pattern of dermal bone in the skull roof and cheek cover; (3) simple jaws, with only two or three pairs of bony tooth plates; (4) a special type of opercular suspension, comprising a dermal submarginal plate connected directly to the braincase, via cartilage of presumed hyoid arch derivation. In this case, the opercular bone-cover that protects the branchial arches is suspended, not by a hyomandibular bone, but by another skeletal structure; (5) a special hard tissue, semidentine, in the dermal bones.

Dermal bones are the bones formed as skin derivatives, and usually are more superficial regarding to the body structure. Semidentine is a kind of dentine that has a strange, "drop"-like, general morphology of its constituent cells. The reader is referred, at this point, to his/her own teeth, which are covered by enamel (the hardest material produced by the vertebrate body). Underlying the enamel, dentine appears (instead of semidentine), and dental bone with a pulp cavity and roots. So, dentine is a very hard kind of bone, and semidentine may either be a kind of dentine only found in †placoderms, or a misleading name for an analogous tissue, similar but of different evolutionary origin, and that maybe should have a different name in order to better indicate the monophyly of †Placodermi.

We note that the first synapomorphy of the †placoderms indeed could have been divided into two. The ring-like shoulder-girdle encircling the trunk, not necessarily forms or produces an articulation with the skull. The neckjoint mentioned in the first synapomorphy, is formed exactly at the same time in evolution as the origin of the shoulder-girdle. The neckjoint could have appeared a little later in evolution, and future fossil findings may show this. Of course, both features may have only existed together (the ring-like shoulder-girdle and its neckjoint), as interpreted by the authors. However, scientifically it would not be ill-intentioned to separate both, in a way that this hypothesis now would count with six instead of five synapomorphies to ground the monophyly of the †placoderms, indeed of a group a little less inclusive than the †placoderms. Alternatively, to reinforce the monophyly of †placoderms, it could be suggested by the authors that a stronger weight would have to be given to the first synapomorphy, because it is (and really is) a character complex, that is, a combination of features that arose only once in evolution. We are discussing this because the final decision on a rival hypothesis is made by means of a parsimony analysis and each evolutionary step obviously counts. We are pointing out that the monophyly of †Placodermi could have been presented in a stronger form in the parsimony analysis. Currently, the monophyly and the polyphyly of †placoderms are in debate (*e.g.* Goujet and Young 2004, Nelson 2006, Burrow and Turner 2010, Zhu *et al.* 2013, Brazeau and Friedman 2014, Dupret *et al.* 2014, Long *et al.* 2015, Qiao *et al.* 2016, King *et al.* 2016). Not all the authors use at least three sequencial outgroups, as should be the case (*e.g.* Maddison *et al.* 1984, Wiley *et al.* 1991: 31) and such less recommended methodological procedures may result in phylogenetic incongruences.

Brazeau (2009) and Burrow and Turner (2010) pointed out that missing data influence the interpretations on the phyletic nature of †placoderms and †acanthodians.

Long (2011: 60) states that there are some 240 known genera of †placoderms. He recognizes seven "major" orders, that is, subgroups of †placoderms. At this point we start discussing whether or not this subdivision of 240 genera into more than one subdivision makes any difference for our argument on the monophyly of terrestrial vertebrates. What kind of impact have the "seven" subgroups accepted by Long (2011)? Here an aside is necessary: the four to six subgroups found previously in the recent phylogenies mentioned above were elaborated using "selected" taxa, whereas Long (2011) did not implement himself a phylogenetic analysis, but instead searched for subgroups of †Placodermi, studying the pertinent literature on the matter. Therefore, he had the opportunity of mentioning all subgroups he could recover from the literature and that he personally agrees with. Therefore, seven, six, four subgroups are not "permanent" numbers for us to have in mind, but are simply numbers of possible ways to group a number of items. In other words, let's permit our minds to work on four to seven possible subgroups of †placoderms.

But what do these groups tell us? If we had only one group for 240 genera, what is the difference? A monophyletic †placoderm means the existence of a single evolutionary group. A paraphyletic †placoderm assemblage means several unrelated groups, each progressively evolutionarily closer to the remaining gnathostomes. A single group of †placoderms mean that all of the species were derived from a single one, and therefore the genetic basis is the same for all. The reason why Long (2011) found seven subgroups does not imply that the genetic basis is not the same for all, but that several different evolutionary lines are recognizable as separate from each other, that is, the differences between them are strong enough to be notable. As a result, these subgroups may be of reasonable different morphologies, distinct ecological roles, and possibly also implying alternative physiological characteristics, as well as unique behaviors. As will be mentioned in Chapter 5 for lungs, it is possible that at least one genus of †placoderm (†*Bothriolepis*) has lungs or a lung-like feature (but see Goujet 2011, for an alternative interpretation). Lungs are very important features for land life (although a family of living salamanders do not have them). The presence of lungs might be a leading feature for terrestriality. On the other hand, Long (2011: 61) mentions the work of Brazeau (2009), who found that some †placoderms (including *Bothriolepis*) do not have pelvic fins and a girdle. The lack of these fins and girdle is one of the arguments for proposing the paraphyly of the †Placodermi. Therefore, we have a †placoderm genus that possibly had lung(s), but did not present pelvic fins (which are homologous to the posterior legs of tetrapods). This situation shows us that some features may represent a new avenue of evolutionary possibilities and others may represent dead evolutionary ends in an "evolutionary avenue" (a taxonomic lineage). The wings of

†pterodactyls, birds and bats are different from each other, but these animals fly or flew. Therefore, different "avenues" were taken to reach the same "condition" (flight). This "movement" is sometimes called evolutionary convergence, but some basic features are needed to reach the same function. These features may be different in each "evolutionary avenue" (a lineage), but they are somehow necessary in each evolutionary lineage for a given function. In the case of flying vertebrates, the anterior pair of locomotor appendages is obviously necessary, as well as its relative independence from the posterior pair, in use or movements, and maybe other pre-existing obligatory prerequisites.

The †placoderm subgroups (no matter if four or seven) have different attributes, and that is why we can tell them apart. These attributes may give evolutionary opportunities for different life styles. Some of them are flat-bodied, similarly to rays and skates. Others are so similar to ratfishes (chimaeras), that they were, some time ago, thought to have originated these last ones, but have a very different life style from the flat-bodied †placoderms. The †*Bothriolepis*-group had pectoral fins with a hinged dermal skeleton that would resemble the legs of a crab; they are thought to be bottom living vertebrates and maybe could "walk" inside water bodies in shallow waters close to dry land with these strong pair of leg-like fins. Other †placoderms had more elongate, almost tubular bodies, and may have been more agile aquatic animals. These differences in body morphology may represent different possible evolutionary ways for several kinds of life styles. For instance, †pterosaurs and birds are both subgroups of Archosauromorpha, animals that include dinosaurs and crocodyles, and possibly had some kind of bipedalism in the origins of the referred large group. Bipedalism, therefore, let the anterior legs free for several uses, including the emergence of wings or wing-like features. Equivalent possibilities (not for flight, of course) may be regarded for each of the different †placoderm subgroups, because their morphologies are different enough to permit different evolutionary developments.

Now, what are the evolutionary implications of considering †placoderms to be monophyletic or paraphyletic assemblages? The evolutionary meaning of a paraphyletic assemblage is that its subgroups are not derived from the same and exclusive DNA at their phyletic origin, at the same moment in evolution. Instead, all †placoderm subgroups share a DNA origin or foundation that is also the DNA shared by all other gnathostomes, including human beings. From gnathostome DNA origin, step-by-step events separate subgroups of what we now call †placoderms. Each of these subgroups are now independent from the remaining subgroups, being a lineage of gnathostomes, instead of †placoderms, that, as a paraphyletic group, does not exist in evolutionary terms. Each group of †placoderms is now considered as an evolutionary lineage of its own. Therefore, after the first, second, third, etc., subgroups of "†placoderms" emerged, they progressively have more to do (are more closely related to) with the remainder of the gnathostomes than to the previously emerged and separated "†placoderms". Each "†placoderm" subgroup is now an independent evolutionary lineage. There is no exclusive DNA at the origin of "†placoderms", because this group is not considered to be a monophyletic group anymore. Therefore, the evolutionary possibilities of each evolutionary "way", or lineage, are completely distinct among each other, if a paraphyletic "†placoderm" is to be considered. Therefore, different lineages are progressively split aside in evolution, emerging as independent sets of evolving DNA. The evolutionary possibilities of each subgroup of a paraphyletic assemblage are therefore detached from each other and do not form an evolutionary unit anymore. The sister group of the remaining paraphyletic "†placoderm" has nothing to do with the remaining †placoderms, which means that there are completely different evolutionary ways for its DNA to evolve and to generate a different life style. Thus, four to seven different evolutionary lineages, not correlated to each other, emerge basally in Gnathostomata, instead of a single one (that would be the case whether †Placodermi are considered as monophyletic). Each splitting event means that "the new taxonomic group does not have anything else to do" with the previously split "†placoderm" groups. For the adaptationist viewpoint, it means one more evolutionary lineage to be tinkered by natural selection (indeed, four to seven more evolutionary lineages). Several more evolutionary lines with separated DNA, that have nothing to do anymore with the remaining subgroup (of "†placoderms", leading to gnathostomes), that in turn, split several times subsequently, leading to more nonrelated sets of DNA, ready to be refined by natural selection. According to Long (2011: 61), all these †placoderm groups appeared at the start of the Devonian, with two subgroups known from the Middle Silurian. Most of these subgroups survived to the end of the Devonian (Long 2011: 61). Therefore, if †Placodermi is a paraphyletic group, natural selection had completely isolated sets of DNA to work out toward terrestriality, not a single one, if †Placodermi is considered to be monophyletic. Thus, the paraphyletic nature of †Placodermi only reinforces our argument on the single origin for terrestriality in vertebrates as an argument against adaptationism, because more groups of gnathostomes, non-related evolutionarily among each other, were available for natural selection to work on toward terrestriality, throughout the mid-Silurian and Devonian periods. None of them led to any known terrestrial vertebrates, living or fossil.

The case of the †Acanthodii seems more complex to analyze. The †acanthodians, with their large eyes and small nasal capsules, were probably mid- and surface-water feeders. Many were microphagous, while others, especially the †ischnacanthiforms, ate fishes and invertebrates (Nelson 2006: 84). In the Silurian they were marine, but in the Devonian they were mainly freshwater (Nelson 1976). The Upper Carboniferous (Mississipian) to Permian *Acanthodes* were distributed in both freshwater and marine environments (Nelson 2006: 84). They reached a maximum length of 2.5 m, but usually measured less than 20 cm (Nelson 2006: 84). A variety of views have been expressed about †acanthodian relationships (Nelson 1976: 51). This author presents the following historical account on them: Watson (1937) reviewed the group and "felt" (quotation marks of ours) that they were the most primitive known gnathostomes. He placed them in the Aphetohyoidea, along with several other groups, which had equal rank as the Pisces (Nelson 1976). In many classifications of the 1930s to the 1950s, they were placed in the class †Placodermi (Nelson 1976). Berg (1940 *ap.* Nelson 1976) recognized †acanthodians in their own class and placed them immediately before his class Elasmobranchii. Romer (1966) provisionally considered them as the most primitive subclass of the Osteichthyes, because of certain resemblances to the actinopterygians. But Nelson (1968a *in* Nelson 1976) "felt" (again, quotations marks are ours) that their closest relative was with elasmobranchiomorphs. Jarvik (1968b: 517 *ap.* Nelson 1976) favored placing them between the Chondrichthyes and Osteichthyes. The view of Moy-Thomas and Miles (1971 *ap.* Nelson 1976), that they are more closely related to the latter, is accepted by Nelson (1976) as the most probable hypothesis. Finally, Nelson (1976) recognizes three orders, †Acanthodiformes, †Ischnacanthiformes and †Climatiformes, this last one with two suborders. Our *fig.* 1-2 is based on the view of Miles (1971 *ap.* Nelson 1976). This historical account of Nelson (1976) for †Acanthodii is very interesting, as it shows first that the position of the taxon varied strongly between basal gnathostomes, and second, that the group always maintained its integrity. In cladistic terms, we would say that it was considered to represent a monophyletic group.

Evolutionary interpretations for the position of †Acanthodii continued to vary. Forey (1980 *ap.* Nelson 1984: 68) returned to the proposal of Watson, but now postulates, in cladistic terms, that †acanthodians are the most "primitive" gnathostome group (quotation marks are ours), sister-group to the others. A more strictly cladistic approach starts to emerge in the literature at this moment of the history of the taxon. Nelson (1984) recognizes again three orders: †Climatiformes, †Ischnacanthiformes, and †Acanthodiformes. Brazeau and Friedman (2014) present a detailed historical account on the matter.

As shown in Hanke and Wilson (2004), in a cladistic analysis of the group, †acanthodians are more diverse than previously appreciated, and the conventional three-order classification is likely an oversimplification, according to Nelson (2006: 84). Several new genera and a family have been described at the time (Nelson 2006: 84). New unstable changes were then proposed, according to him. Therefore, he maintains the three orders of his previous edition (Nelson 1984).

The strict consensus tree of Burrow *et al.* (2016) shows a monophyletic †Acanthodii that in turn is simply a large polytomy, including a monophyletic Chondrichthyes. This large polytomy does not explain much, bringing doubts instead. The polytomy may simply represent that †Acanthodii is the sister group of Chondrichthyes instead of Osteichthyes, as in our *fig.* 1-2. Basically, no new interpretation regarding the monophyly of terrestrial vertebrates emerges from this possibility. However, other possibilities are opened when a polytomy remains in a cladogram, and it is difficult to conceive all of them. Still, if †Acanthodii is a monophyletic group that includes the cartilaginous chondrichthyan "fishes", then a number of evolutionary "ways" for terrestriality are possible, that is, †acanthodian lineages (including Chondrichthyes), is much larger than what our *fig.* 1-2 shows. Differently from the †placoderm paraphyletic condition, where the groups have nothing to do with each other in evolutionary terms, because †"placoderms" follow a sequential splitting that makes some groups progressively closer to the remaining gnathostomes in evolutionary and biological terms, the case of †acanthodians still is of a monophyletic group (that now includes the chondrichthyans). This means that there is a unity in DNA of the whole group, because †Acanthodii is paraphyletic only in the sense that now it includes Chondrichthyes. Otherwise, †Acanthodii continues to be a single, whole, integrate evolutionary lineage that only includes one more subgroup, the chondrichthyans. Regarding †acanthodian paraphyly, we do not figure out a clear possible difference for our original interpretation of terrestrial vertebrates, in clear contrast to that depicted from †placoderm paraphyly.

King *et al.* (2016) also present †Acanthodii as including the chondrichthyans.

Brazeau (2009) brings a very different picture of relationships within †acanthodians. His *fig.* 3 is a strict consensus tree of 2,904 shortest trees from the global analysis. From this tree it is possible to see that Brazeau's (2009) †acanthodian paraphyly range widely, that is, a group of genera long known to be †acanthodians (such as †*Acanthodes*, †*Homalacanthus*, †*Ischnacanthus*, and others) lie as the sister group of Osteichthyes, while other †acanthodian genera, such as the also long known †*Climatius*, lie closer to, or inside, the Chondrichthyes. That is, the paraphyly of †Acanthodii found by Brazeau (2009) leads to a similar interpretation as that we developed for the paraphyly of the †Placodermi. At least two diverging groups of †acanthodians are depicted: those closer to Osteichthyans, and those closer to Chondrichthyans. Other groups remain as *incertae sedis*, that is, they have unresolved phylogenetic positions. We are not going to make further comments on them, because they might simply be closer to chondrichthyans, in which case they would have no new or different meanings. Brazeau and Friedman (2014: 812, *fig.* 13A) is not much different from Brazeau [2009, *fig.* 3] regarding the distribution of †acanthodian genera in the cladogram. As Brazeau (2009, *fig.* 3), Brazeau and Friedman (2014) indeed only divide †acanthodians into the lineages that are closer to Osteichthyes (as in our *fig.* 1-2), and those that are closer to Chondrichthyes. Interestingly, Janvier (1996: 331, *fig.* 9-1) already presents a widely "divided" †Acanthodii, with "some †acanthodians" lying closer to chondrichthyans and some others lying closer to osteichthyans, reminding us of Brazeau's (2009) phylogeny. Janvier's phylogeny differs for not bringing genus names, only the term "some" in regard to groups of genera or species. Although far from the impacting †placoderm paraphyly, the paraphyly Brazeau (2009) found for †Acanthodii enhances our original proposition on the monophyly of terrestrial vertebrates as an argument against adaptationism and in favor of exaptations at the origin of this terrestriality. If these propositions were to be followed, at least one new terminal would have to be placed in our *fig.* 1-2 (closer to osteichthyans), and that represents a new evolutionary lineage (in our referred *fig.*), from which at least one terrestrial group of vertebrates could have emerged. But such an emergence has not occurred, so far as known.

CHAPTER 5

Method Based on Cladistics: New Perspectives

Phylogenetic systematics or Cladistics is a tool that permits a clear proposal for or against natural selection as the main explanation for certain evolutionary processes, as was discussed in the previous Chapter. In that Chapter, we saw how monophyly and polyphyly may sustain hypotheses of macroevolution and microevolution, respectively. As we will see here, each character may also be evaluated individually, on the basis of cladograms, further refining argumentations in favor of, or against one or the other evolutionary process.

Blackburn (2002) developed a procedure that permits us to discern evolutionary novelties (apomorphies) that are adaptive from those that are exaptive. According to Blackburn (2002: 507), a crucial difference between adaptation and exaptation refers to the chronological sequence of structural modifications when these are overlapped with the chronological sequence of functional changes. The moment of the evolution of a structure is inferred from its taxonomic distribution (Brooks and McLennan 1991, *ap.* Blackburn 2002: 508). This is equivalent to saying that the moment of appearance of an evolutionary novelty corresponds to the point in the cladogram at which an apomorphy arises. Thus, inserting apomorphies and, independently, their functions in the same cladogram reveals possible adaptations when function anticipates or accompanies structural characters. When a structure appears before, and only later (in the cladogram) the aimed function appears, exaptation is suggested (Blackburn 2002: 508). A structure that anticipates its function is a sufficient criterium for the recognition of an exaptation. However, a function that anticipates or accompanies the evolution of a particular structure is a necessary, but not sufficient, criterium for the recognition of an adaptation. Other criteria (such as those discussed in Andrews *et al.* 2002) must be given to support this idea (Blackburn 2002: 508). In practice, a cladistic analysis is useful particularly when a robust cladogram may be constructed on the basis of characters that vary in their structural and functional attributes (Blackburn 2002: 508).

In order to clearly understand the method of Blackburn (2002), it is necessary that we insert taxa, apomorphies, functions, and effects in a same cladogram. Below, we present data from our article on the subject (Triques and Christoffersen 2015). In the case of the origin of terrestriality of vertebrates, this implies a very large cladogram, which includes Actinopterygii and Sarcopterygii, with an enormous detailing of the Sarcopterygii. Sarcopterygii comprise

 | DOI:10.1163/9789004684911_006

a taxon having a very large fossil diversity. It is precisely in these fossils that one encounters the origin of most (but not all) anatomical characters used by vertebrates on land. Some of these traits are present also in Actinopterygii, as we shall see below. It would thus be impossible to present a complete cladogram on a single page, and furthermore including apomorphies, functions, and effects, as the method requires. Consequently, it has been necessary to greatly simplify the cladogram of the Sarcopterygii, in order to attain a result which is sufficiently clear and didactic.

To see the full picture, all information must be on one page for the reader. Looking at the published cladograms, none is such that it includes all the necessary terminals we need for our argumentation. We needed one single phylogeny to include the indispensable terminals, including space for the concerned apomorphies, their functions, and their effects. Therefore, we had to combine and simplify cladograms into a single image for the reader. McLennan (2008: 252–258, *fig.* 3) came to a similar conclusion studying other features or depicting characters in a somewhat different manner, but she preferred "co-option" instead of exaptation. She correlated exaptation to preadaptation, stating that Gould and Vrba eliminated the specter of teleological terminology by replacing "preadaptation" with the concept of "exaptation" (McLennan, 2008: 249). Thus, we worked unaware of her results, analyzed other characters, and in a different way, with other arguments and perceptions. Due to the first author of this book's failure to perceive McLennan's (2008) work as a criticism of adaptationism until the final completion of this book, we ended up developing our own reasoning. Perhaps this book would not exist if we had already known about McLennan's work (2008). Therefore, we decided to maintain the original text here, since all our proposals are effectively original. Agosta and Brooks (2020: 206–208, *figs.* 9-1, 9-2) also utilize the term co-option, modified McLennan (2008: *fig.* 3) in a way that is closer to that of Triques and Christoffersen (2015: *fig.* 1, 2018, *fig.* 1) and consequently to our *fig.* 5-1.

Also, to implement the procedures suggested by Blackburn (2002) with regard to the structural and functional characters associated with the origin of the terrestrial vertebrates, it has thus been necessary to unite and fuse some cladograms present in the pertinent literature, simplifying the remaining whole, by pruning selectively several terminals, as clarified below. The result of this procedure is expressed in *fig.* 5-1.

Apomorphic structures associated with the conquest of land were inserted into the cladogram, as well as their correlated functions, independently of each other. From this scheme, it is possible to discern adaptive from exaptive apomorphies for terrestriality. Finally, we will discuss if it is appropriate to

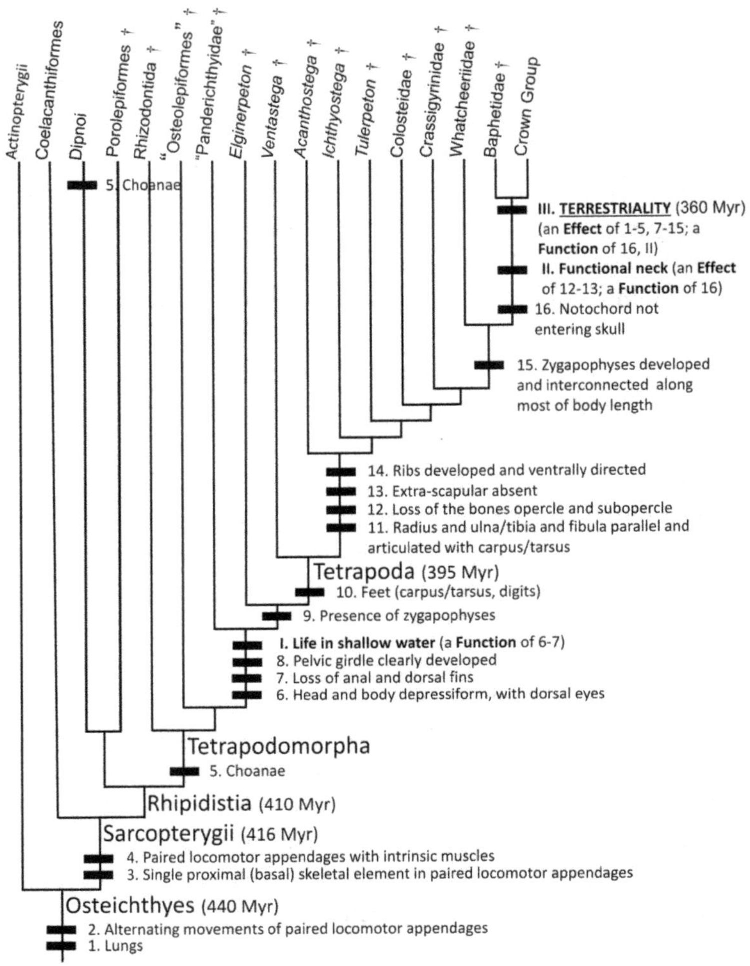

FIGURE 5-1 Cladogram of Osteichthyes with selected taxa, illustrating the origin of terrestriality in Tetrapoda (vertebrates with feet). Characters are numbered with Arabics (1–16) and lifestyles are numbered with capital Romans (I–III). Numerals (Arabic and Roman) connected by an hyphen in the cladogram signify simply that more than one character or lifestyle appeared at that level of the cladogram, not that they are connected in any way, or ordered in any direction. Minimal ages are indicated for selected taxa and for the event of terrestriality, in millions of years ago (Myr), in brackets or separated by a hyphen. The height of an internode (the vertical bar where frequently characters are indicated) only expresses the number of apomorphies (characters discussed in this book), effects, functions, lifestyles and names of taxa used in this study (a higher or shorter height is simply necessary to accommodate the information in a particular internode) and, consequently, no other attribute should be associated with differences in height (or length) of the internodes. The crown group is the monophyletic group containing the terrestrial tetrapods with living representatives (Batrachomorpha + Reptilomorpha + Lepospondyli), more recently termed Neotetrapoda by some authors.
MODIFIED FROM TRIQUES AND CHRISTOFFERSEN (2015, 2018, PUBLISHED WITH PERMISSION). SEE TEXT FOR MORE INFORMATION.

correlate these characters with "adaptive scenarios". Our aim was not to discuss apomorphies in the cladogram exhaustively. We only intend to add representative apomorphies to clearly illustrate aspects related to the occurrence of adaptations and/or exaptations in the origin of the Tetrapoda and their conquest of the terrestrial environment. In other words, we ask at which moment in Sarcopterygii did the appearance of fossil characters related to terrestriality appear, and which was the main evolutionary process involved, micro- or macroevolution.

Along the text, we refer to a taxon or taxa that split toward the base of the cladogram as the "basal taxon" (or basal taxa, depending on the case), relative to the more apical taxon which we are dealing with at the moment. The term "basal" does not imply that the taxon is "primitive", "inferior" or "worse" than the others, only that it is the sister group of another more diversified group that becomes subsequently subdivided into increasingly subordinate taxa, resulting in a final asymmetrical cladogram (which means that there are more terminals on one side of the cladogram than on the other side). For example, in *fig.* 5-1, Coelacanthiformes and Dipnoi + †Porolepiformes (composing the taxon Dipnomorpha) are more basal in the cladogram in relation to the remaining sarcopterygians.

Sarcopterygii are defined herein as in Benton (2005: 69). We include taxa with paired pectoral and pelvic muscular limbs, having a substantial number of internal bones (among other synapomorphies). We also include the presence of a single proximal skeletal element (that is, at the base of each fin, linking the sustaining girdle with the body – which in the arm represents the humerus and in the leg represents the femur), and which is called the "stylopod". However, it should be mentioned that Zhu and Yu (2009) found three or four skeletal elements in representatives of basal sarcopterygians (in two genera, so far) and considered that only the "crown group" of Sarcopterygii (the subgroup of Sarcopterygii with living representatives and their related fossil members) would be characterized by the presence of a single skeletal supporting element. For the sake of simplicity and clarity, we here consider the occurrence of a single basal element in the paired locomotor limbs as a synapomorphy of Sarcopterygii.

We combine and simplify three cladograms of Benton (2005: 68, 80, 98, tables 3.6 and 4.1), ommiting †*Obruchevichthys* (the sister group of †*Elginerpeton*), †*Metaxygnathus* (the sister group of †*Ventastega*) and the taxa included in Batrachomorpha, Lepospondyli, and Reptilomorpha (all belonging to the crown group, the last terminal in the illustrated cladogram). On the other hand, we include Rhizodontida as the sister group of the remaining tetrapodomorphs (*fig.* 5-1). At this time, we stress that the method based on a cladogram used in this work was benefited by simplifications, because this reduced cladogram was large enough (from a taxonomic perspective) to include, in a same figure, all studied apomorphies, together with their functions and/or effects. We have thus chosen and simplified cladograms found in Benton (2005) with this aim in mind.

Apomorphies were included in the cladogram of *fig.* 5-1 following the original cladograms. "Life styles" in shallow water and terrestriality, as well as the presence of a functional neck, were included in the cladogram based on information from the literature mentioned along the book. The interpretation that these life styles are a function or effect of certain apomorphies was made based on the method of Blackburn (2002) itself, and were also included in the cladogram (*fig.* 5-1).

Several cladograms have been published for the Sarcopterygii, some of which place the Tetrapoda as the sister group of the remaining sarcopterygians (*e.g.* Chang 1991 *ap.* Benton 2005) or of the basal groups (Romer 1966: 47, *fig.* 67). However, when †Panderichthyidae, †Elpistostegidae or "†Panderichthyida" are included in the cladistic analysis, Tetrapoda appear as their sister group, jointly emerging as the most derived group of sarcopterygians. This is central to our discussion of adaptation/exaptation in the origin of the Tetrapoda: the more detailed the cladogram, the more derived is the position of "†Panderichthyidae" + Tetrapoda in the cladogram. Thus, it is not necessary to use the most recent or the most complete cladogram from the literature, in order to discuss characters as we plan. We follow Clack (2012) in her belief that "at this time, it is difficult to imagine that opinion will change much from the idea that tetrapods descend from a lineage of tetrapodomorphs, including fish such as †*Tiktaalik*, †*Panderichthys*, †*Eusthenopteron* and †*Osteolepis*". "†Panderichthyidae" is written here between quotation marks to indicate its paraphyletic nature, as †*Panderichthys* is placed more basally, while the remaining genera are more closely related to Tetrapoda (*e.g.* Clack 2012: 60, *fig.* 3-2; Niedzwiedzki *et al.* 2010: 43). In the quest for simplicity, however, we do not divide this group further into its diverse subgroups. †Baphetidae is usually positioned as the sister group to the crown group (*i.e.* the group of organisms directly related to modern tetrapods) or sometimes one step more distant from this group (*e.g.* Laurin 2010: 162, *fig.* 7-1). These different interpretations do not change our final results, as will be discussed at the appropriate place in the text. The true phylogenetic position of †*Crassigyrinus* is still under discussion (Clack 2012: 287), but most studies place this genus outside the crown group. "†Osteolepiformes" (*figs.* 5-2 and 5-3) include several taxa that may be progressively more closely related to Tetrapoda (*e.g.* Clouthier and Ahlberg 1996: 464, *fig.* 3), and for this reason it is cited within quotation marks. We prefer the term "Coelacanthiformes", rather than "Actinistia" (*fig.* 5-4), because it appears more frequently in the popular literature. †Onychodontida were omitted for being insufficiently known regarding some key-characters, such as choanae (*e.g.* Clouthier and Ahlberg 1996: 465), structures which we consider in *figs.* 5-5 and 5-6. As this group possibly represents a basal lineage of rhipidistians (Rhipidistia), its absence in the cladogram does not alter our reasoning. Actinopterygii are the sister group of Sarcopterygii (*e.g.* Zhu and Yu 2009; Lauder and Liem 1983, among many others) (*e.g. figs.* 1-1B, 1-2, 1-3, 5-1, 5-7 and 5-8), both groups composing the Osteichthyes. Dipnoi include fossil forms (*fig.* 5-9) and living animals (*figs.* 2-1D and 5-10), including the South American lungfish.

Actinopterygii are a very large taxon which we are unable to include in depth in this book. There are about 30 thousand valid living species (Friedman and Giles 2016). This is a gigantic number for a subgroup of vertebrates, being in fact the largest in number of species. In *fig.* 1-1B we have a tarpon, a popular game fish in the U. S. Actinopterygii have several fossil groups, which we shall not mention. A simplified general view of Actinopterygii was presented at the Chapter 1, before dealing with the skeleton, including Teleostei and its subgroups. Its early evolution was studied in Chapter 4, in order to provide some new arguments against adaptationism and new arguments in favor of macroevolution in the passage of vertebrates from water to land. To deal with all this diversity in Actinopterygii would take the focus away from the Sarcopterygii in the present Chapter, in which the truly terrestrial groups are included. Most importantly, dealing more extensively with the actinopterygians would dilute the main focus of the book, which is the discussion of arguments favoring macroevolution in the transition from water to land in the vertebrates.

Coelacanthiformes are presented in *figs.* 5-1 and 1-4 as the sister group of the remaining sarcopterygians, and includes the living genus *Latimeria* (*fig.* 2-2A), which has two living species. In the past, the diversity of this group was quite large (*e.g.* Janvier 1996). Living species are known from deep waters over the continental platform of the Indian Ocean and, maybe for this reason, its lungs are filled with fat, changing its respiratory function to a hydrostatic role (adjusting the total density of the body to approach that of the water at given depths). They feed mainly on cephalopods, have conical teeth, the body coloration is dark blue with white spots (*Latimeria chalumnae*), and they have internal fecundation, but a copulatory organ is not known.

Dipnoi include the South American lungfish (*Lepidosiren paradoxa*). They occur in northern Argentina, the Brazilian Pantanal, the Amazon, and maybe also occur in Surinam. They are called lungfish, because of the presence of functional lungs. Both the South American representative and the African species (genus *Protopterus*, *fig.* 2-2D) are capable of resisting drought,

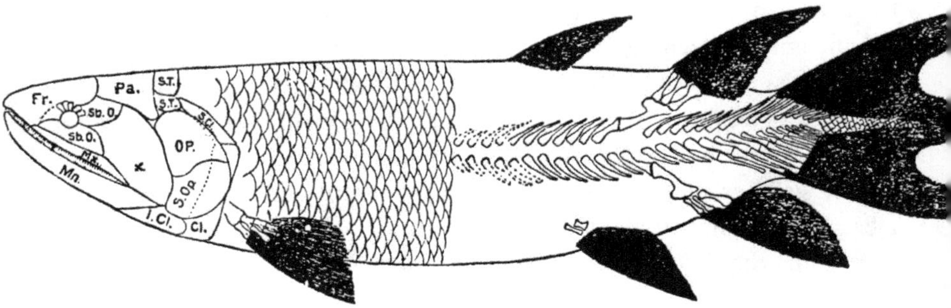

FIGURE 5-2 Reconstruction in lateral view of the "†osteolepiform" (Tristichopteridae, Sarcopterygii) †*Eusthenopteron foordi*. Scales represented only on anterior part of post-cranial region. Features are cl, clavicle; fr, frontal; i.cl, infraclavicle; mn, mandible; mx, maxilla; op, operc pa, parietal; s.cl, supraclavicle; s.op, subopercle; s.t, supratemporal; sb.o, suborbital; x, check-plate

FROM WOODWARD (1898) IMAGE IN THE PUBLIC DOMAIN

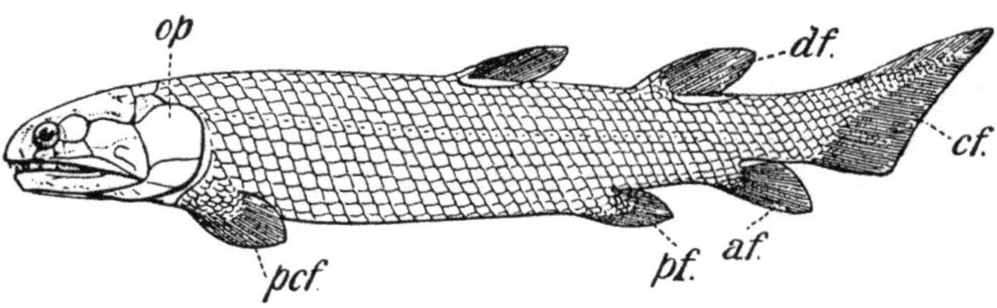

FIGURE 5-3 Reconstruction of the "†osteolepiform" †*Osteolepis macrolepidotus* (Sarcopterygii) in lateral view. Indicated features are af, anal fin; cf, caudal fin; df, posterior dorsal fin; pcf, pectoral fin; pf, pelvic or ventral fin; op, opercle

FROM GOODRICH (1909). IMAGE IN the PUBLIC DOMAIN

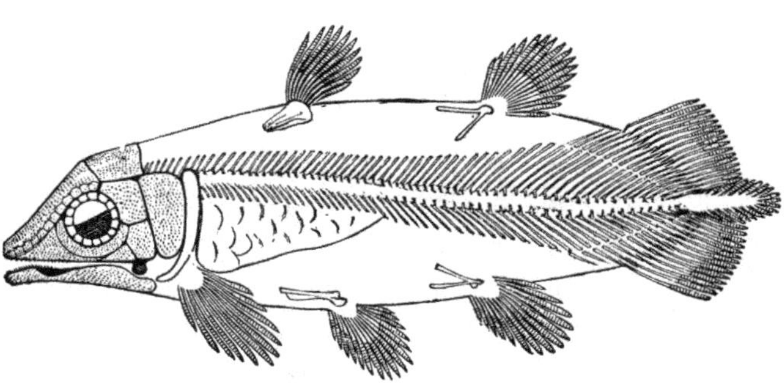

FIGURE 5-4 Lateral view of skeleton of the fossil coelacanth (Actinistia, Sarcopterygii) †*Undina* [=*Holophagus*] *gulo*. Scales and superclavicula omitted. The extent of the ossified air-bladder is indicated beneath the notochordal axis in the abdominal region

FROM WOODWARD (1898). IMAGE IN THE PUBLIC DOMAIN

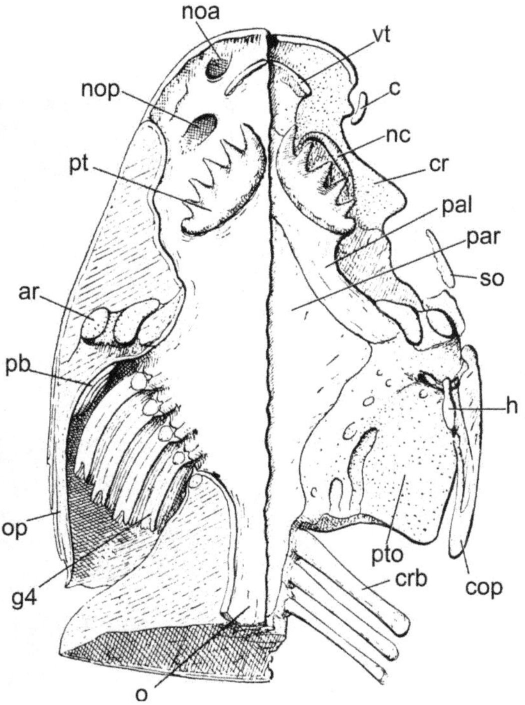

FIGURE 5-5
Ventral view of roof of the mouth, or palatum of the living Australian lung fish *Neoceratodus forsteri* (Dipnoi, Sarcopterygii). Features are as follows: ar, articular facet of quadrate; c, labial cartilage; cop, opercular cartilage; cr, cartilaginous cranium; crb, cranial (anterior) rib; g4, fourth gill arch; h, vestigial hyomandibular; nc, nasal cartilage; noa, anterior "external" nostril; nop, posterior internal nostril (the choana); o, oesophagus; op, operculum; pal, palatopterygoid bone; par, parasphenoid; pb, pseudobranch; pt, palatine tooth; pto, pterotic; so, suborbital; vt, vomerine tooth
MODIFIED FROM GOODRICH (1909). IMAGE IN THE PUBLIC DOMAIN

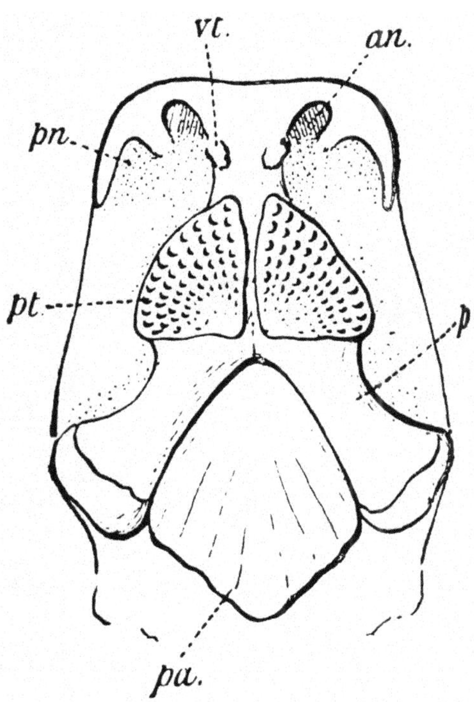

FIGURE 5-6 Reconstruction of the roof of the mouth, or palatum, of the fossil lung fish †*Dipterus sp.* (Dipnoi, Sarcopterygii), with the following structures: an., anterior nostril; p., fused palatine and pterigoid bones; pa., parasphenoid; pn., posterior nostril or choana; pt, palatine teeth; vt., vomerian tooth
FROM GOODRICH (1909). IMAGE IN THE PUBLIC DOMAIN

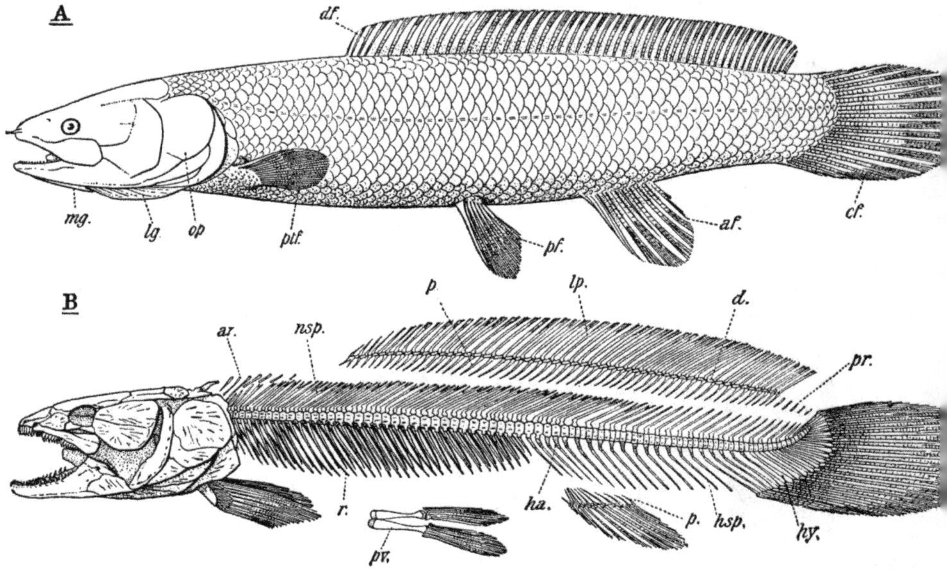

FIGURE 5-7 *Amia calva* (Halecomorphi, Actinopterygii), a North-American freshwater fish ecologica similar to the trahiras of South America, in lateral view. A. External anatomy. B. Skeleton Features are labeled as follows: af, anal fin; ar, anterior dorsal radials (supraneurals); cf, caudal fin; d, distal segment of dorsal radial (distal radial); df, dorsal fin; ha, haemal arch hsp, haemal spine; hy, hypural arches; lg, lateral gulars (branchiostegal rays); lp, lepidotr (bony dermal rays); mg, median gular (urohyal element); nsp, neural spine; op, operculu p, proximal segment of radial (proximal radial); pf, pelvic (or ventral) fin; pr, posterior d radials (epurals); ptf, pectoral fin; pv, pelvic bone; r, pleural rib
FROM GOODRICH (1909) IMAGES IN THE PUBLIC DOMAIN

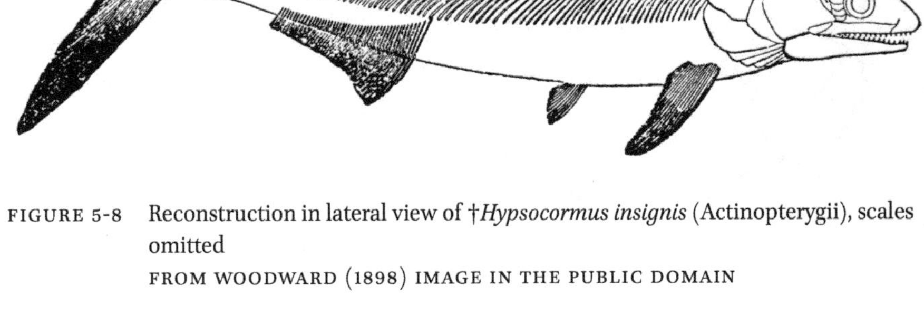

FIGURE 5-8 Reconstruction in lateral view of †*Hypsocormus insignis* (Actinopterygii), scales omitted
FROM WOODWARD (1898) IMAGE IN THE PUBLIC DOMAIN

constructing a burrow in the mud and entering it before the water of the river dries completely. They remain in this burrow, breathing air, consuming their fatty reserves, and even consuming proteins to maintain their metabolism low, when necessary. Dipnoi (*e.g. figs.* 2-1D and 5-9), may be characterized by a buccal apparatus capable of grinding hard food (called "durophagic", *figs.* 5-5 and 5-6), such as the carapace of invertebrates. The information presented here on Dipnoi, Coelacanthiformes, and Actinopterygii may be found in many textbooks (*e.g.* Pough *et al.* 2005), while some were obtained from Lauder and Liem (1983) and Nelson (1994).

Crown group is an expression frequently used by paleontologists to name a monophyletic group containing living representatives, besides the fossils. This expression appears several times in this book and in *fig.* 5-1, to refer to a very large group without going into further details when these extend beyond the scope of this book. Gaffney (1979) presented a phylogenetic analysis supporting the monophyly of Tetrapoda, indicating within this group a sister-group relationship between †*Ichthyostega* and a monophyletic group termed by Gaffney the "Neotetrapoda" (quotation marks of Panchen 1991). Panchen (1991: 112, *fig.* 1) shows that Neotetrapoda include "Temnospondyli" (quotation marks of Panchen) + Lissamphibia, both being the sister groups of †Lepospondyli + "†Anthracosauria" (quotation marks of Panchen) + Amniota, that is, the same group we are terming in this book as the crown group Tetrapoda. Recently some authors (*e.g.* Vorobyeva 2008, Benton 2015: 109–110) have utilized the term Neotetrapoda. However, we consider that "crown group" or "crown-group" highlights the idea of being "the" taxonomic group that includes all extant subgroups of Tetrapoda and their phylogenetically closer fossil relatives. In the definition given by Gaffney (1979), Lissamphibia (*fig.* 1-4) and Amniota (*fig.* 1-4) are the living representatives of Neotetrapoda and Temnospondyli.

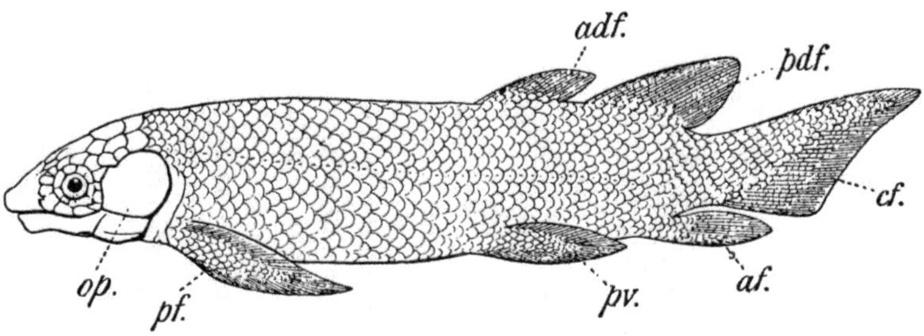

FIGURE 5-9 Reconstruction of the fossil lung fish †*Dipterus valenciennesii* (Dipnoi, Sarcopterygii) in lateral view. Characters are: adf, anterior dorsal fin; af, anal fin; cf, caudal fin; pdf, posterior dorsal fin; op, operculum; pf, pectoral fin; pv, pelvic or ventral fin
FROM GOODRICH (1909). IMAGE IN THE PUBLIC DOMAIN

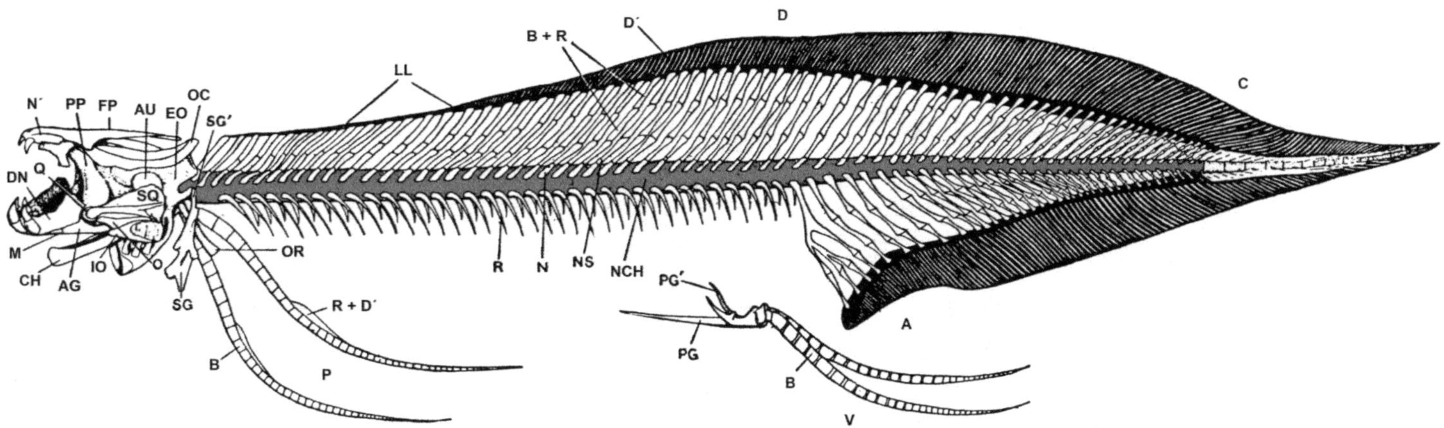

FIGURE 5-10 Lateral view of skeleton of the living African lungfish (Dipnoi, Sarcopterygii) *Protopterus annectens*. Features are as follows: A, anal fin region; AG, angular; AU, auditory capsule; B, cartilaginou axis of fin; B + R, basals and radials; C, caudal fin; CH, ceratohyal; D, dorsal fin region; D′, dermal fin rays; DN, dentary; EO, epiotic; FP, lateral or "fronto-parietal"; IO, subopercular; LL, longitudinal ligament; M, Meckel′s cartilage; N, neural arch; N′, nasal bone; NCH, notochord (in grey); NS, supraneural; O, opercular; OC, occipital crest; OR, occipital rib; P, pectoral fin; PG, pelvic girdle; PG′, dorso-lateral process of PG; PP, palatopterygoid; Q, quadrate; R, rib; R + D′, radial and dermal fin elements; SG, shoulder girdle or pectoral girdle; SG′, dorsal process of shoulder girdle (SG); SQ, squamosal; V, ventral or pelvic fin

MODIFIED FROM DEAN (1895).

†Lepospondyli and †Anthracosauria are their fossil relatives. The quotation marks used by Panchen indicate the current possibility that these may indeed not be monophyletic groups. As Lissamphibia may be part of Temnospondyli, we are not including the symbol "†" in front of it.

1 Criticisms of Adaptationism

Although without intending to be exhaustive, Andrews *et al.* (2002) present a critical review of "adaptationism", focused on evolutionary psychology, but also stressing that the concept of exaptation represents "an exaptation program". According to these authors, adaptationism is a research strategy that aims to identify adaptations and to clarify which selective processes molded these adaptations in the evolutionary past of an organism. Although everyone agrees that organisms have adaptations, adaptationism, as a research strategy, has not enjoyed a consensus within evolutionary biology. During the 70s, adaptationism became the subject of criticisms by paleontologist Stephen Jay Gould and geneticist Richard Lewontin (*e.g.* Lewontin 1978, 1979, Gould and Lewontin 1979). Andrews *et al.* (2002: 489), mentioning Williams (1966), consider that maybe adaptationists use intuitive arguments to show how a particular character responds to evolution. As an example in line with the theme of this book we may cite that Clack (2002: 99–104) identified several scenarios to explain the evolution of limbs with digits, supposedly in intimate connection with the environments of the first tetrapods. For her, it is difficult to dissociate these two components of evolution. Our perspective is closest to that of Gould and Vrba (1982: 11), in which the order and arrangement of the bones in the limbs of tetrapods are exaptations for walking on land. Gould and Vrba (1982: 11) apparently refer to limbs as in Romer and Parsons (1977: 195–196): tri-segmented, with the proximal part containing a single bone (femur in the leg and humerus in the arm), followed by two approximately parallel bones (tibia and fibula in the leg and radius and ulna in the arm) and finally followed by the foot, which includes the wrist. Many changes in shape and muscles are secondary adaptations for terrestrial life (Gould and Vrba 1982: 11). Here we formally explain this "exaptative" view of the evolution of the limb of a tetrapod, implementing the methodological procedure of Blackburn (2002).

The aim of adaptationism is to determine if traits are adaptations (Mayr 1983, *ap.* Andrews *et al.* 2002: 493). To classify a trait as an adaptation is to identify its function (Thornhill 1997 and Williams 1966, *ap.* Andrews *et al.* 2002: 493). And to identify a function is to determine specific selective pressures (if any), at least partially responsible for the evolution of the trait (Andrews *et al.* 2002:

493). Through the years, Gould, Lewontin and their colleagues have argued that adaptationism is not only a fallible method for determining the effect of evolution in general, but also for the understanding of the specific effect of its main concern, selection (Andrews *et al.* 2001: 493). Their criticisms center around a common argument: adaptationism is based on an evolutionary view that exaggerates the power of selection and undervalues the restrictions to selection and the action of other possible evolutionary processes (Andrews *et al.* 2002: 493). They do not deny that selection may be responsible for changes in preexisting characters toward a "workable design". Gould (1997: 57) recognizes that natural selection is the only known cause for producing a "workable design". In other words, characters change in evolution in response to natural selection. And Gould does not deny that natural selection is the primary evolutionary force responsible for evolutionary change (Gould 1984). The problem is that factors other than natural selection may be interpreted by an adaptationist as also representing the results of natural selection (Andrews *et al.* 2002: 493). For us, the origin of tetrapods and their conquest of land is currently linked to all of the following statements: an adaptationist view of the entire evolutionary process, from feet to life on land, as dependent on evolutionary scenarios, and related exaptations have been – in our view – under-evaluated and/or neglected. The point of disagreement between adaptationists and their critics is centered on the corroboration value of the evidence that adaptationists use to classify a trait as an adaptation, which is nothing more than the consistency between adaptationist hypotheses and the evidence for the function of the trait (Andrews *et al.* 2002: 493).

Biologists use the term "trait" to refer to aspects of the phenotype of organisms. However, what qualifies as a trait is not given so directly. A liberal definition would characterize a trait as any aspect of the phenotype which may be discriminated in any way – causes, effects, its appearance, for example (Andrews *et al.* 2002: 490). Behaviors are not traits, because they are not constructed from genes and their products. Behaviors are rather effects of the components, or of the interacting components of the nervous system (*e.g.* an emotional experience) or effects of the nervous system interacting with the skeleto-muscular system (*e.g.* behaviors). However, behaviors and physiological processes are like traits, in the sense that they produce effects (*e.g.* the movement of the hand modifies the environment to create a tool) and these effects are frequently functional (Andrews *et al.* 2002: 490–491). Thus, we consider the alternate movements of the paired fins in sarcopterygians to be a behavior, that may be used to test the selected versus exapted nature of the use of their paired fins (or paired locomotor appendages).

As all aspects of the phenotype are integrated, organisms are not collections of discrete objects (they are not isolated) (Gould and Lewontin 1979 *ap.*

Andrews *et al.* 2002: 490). Genes frequently have pleiotropic effects (*i.e.* a single gene may influence many aspects of the phenotype of an organism) and they frequently interact epistatically with each other (*i.e.* a gene allele from one *locus* may influence the phenotypic expression of an allele in another *locus*).

Biologists interested in how the phenotype of an organism evolves are forced to discriminate aspects of the phenotype (Andrews *et al.* 2002: 490). This matter will be discussed later, with the nature of the paired locomotor appendages of sacopterygians.

Gould and Vrba (1982) were the first to define and discuss the concept of exaptation (Andrews *et al.* 2002: 491). An exaptation is a preexistent trait (*i.e.* a trait that had already appeared, evolved) that acquires a new beneficial effect without being originated by selection for this effect (*i.e.* the trait assumes a new role, but was not "planned" for such an effect, via selection). Because the beneficial effect did not contribute to the evolution (emergence, origin) of the trait, the use for which the trait is exapted is not called a function, but an effect. "Adaptations have functions; exaptations have effects" (Gould and Vrba 1982: 6). Subsequently, we show how exactly most mechanical structures used by tetrapods for living on land appeared much before their use on land and also much before any attempt to "walk" on land. For a trait to become exapted for a new beneficial effect, it must have been acquired without having phenotypically originated by selection for that effect (Andrews *et al.* 2002: 491). This seems to be the case of most mechanical characters mentioned above, which represent information left in the fossil record awaiting interpretation. Soft parts are fossilized more rarely. On the other hand, any coopted effect (an exaptation) probably will not appear perfected for its new use (effect) but will develop secondary adaptations for the new function (Gould and Vrba 1982: 13). In principle, then, primary exaptations and secondary adaptations may be distinguished.

2 Characters

The studied features are presented and discussed below and numbered in *fig.* 5-1 as structural features (1–16), together with the respective life style features, when applicable (Roman numerals I-III), followed by their respective functions (F) and/or effects (E).

Paired locomotor appendages (3–4, 11, III). Sarcopterygii present muscular paired locomotor appendages, which are also provided with an elongated axial skeleton – the so-called paired lobe fins – (*fig.* 5-11). Each lobe, in turn, has a single element contacting the adjacent skeletal girdles (*fig.* 5-12). This

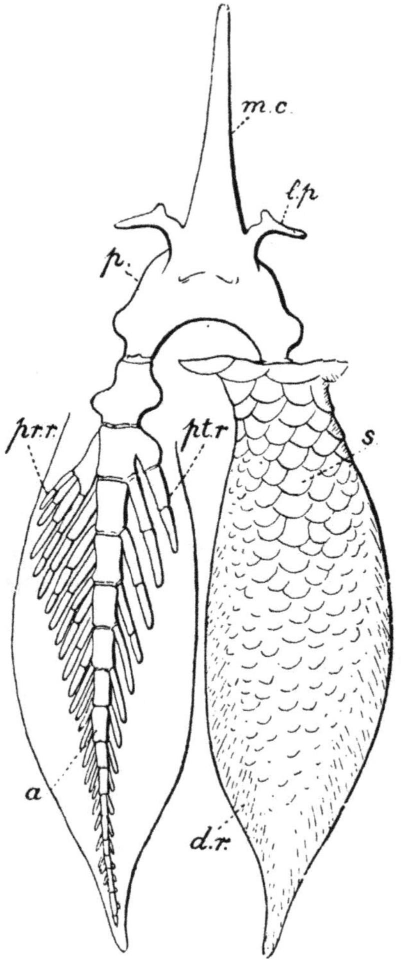

FIGURE 5-11 Ventral view of pelvic (or ventral) fins and pelvic girdle of the living Australian lung fish *Neoceratodus forsteri* (Dipnoi, Sarcopterygii), anterior region upwards; left side represents the fin skeleton and the right side represents the external anatomy. Features are: a, articulated axis of the fin; d.r, dermal rays; l.p, lateral prepubic process; m.c, median epipubic process; p, pelvic girdle; pr.r, post-axial radial; pt.r, preaxial radial; s, scale covering axial region FROM GOODRICH (1909). IMAGE IN THE PUBLIC DOMAIN

combination does not occur in other vertebrates (*fig.* 5-13), with rare exceptions (*fig.* 5-14), which are considered homoplasies for this character of Sarcopterygii. As homoplasies, they are considered to have emerged independently of each other.

Whether or not it is possible, or desirable, to split paired appendages into more than a single synapomorphy for Sarcopterygii is a matter of interest both to cladistic analysis and to evolutionary studies. Independently of the final decision about this, it is possible to discuss the whole set as a single structure. To elevate the body and support weight on and above land, and yet to walk forwards freely, tetrapods use not only four feet but also skeletal struts directed downwards from the girdles: the four (two pairs of) walking limbs. Each limb

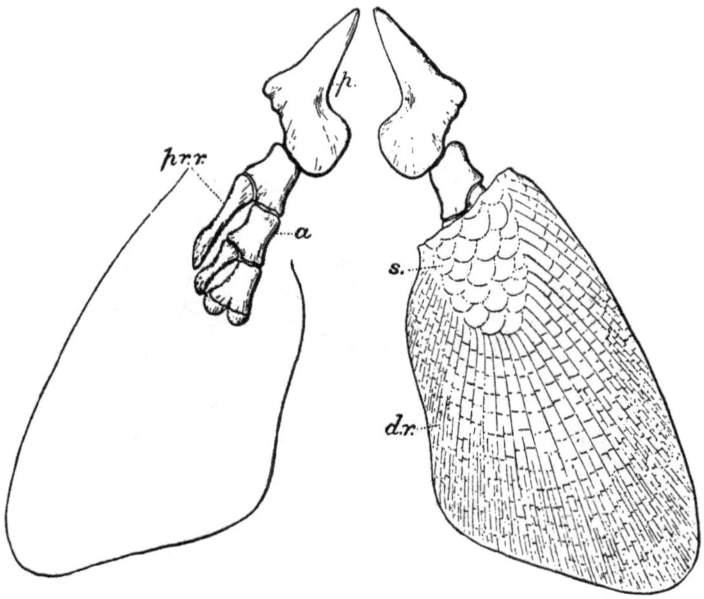

FIGURE 5-12 Ventral view of pelvic (or ventral) fins and pelvic girdle of the "†osteolepiform" †*Eusthenopteron foordi* (Tristichopteridae, Sarcopterygii), anterior region upwards; left image represents the fin skeleton and right image represents the external anatomy, including the scales (s) and dermal rays (d.r.). Other indications: a, articulated skeletal axis of fin; pr.r., post-radial radials; p, pelvic girdle
FROM GOODRICH (1909). IMAGE IN THE PUBLIC DOMAIN

has a long internal axial skeleton, which is in fact responsible for elevating the body above ground. This skeleton is composed of humerus and radius/ulna in the anterior limb and femur and tibia/fibula in the posterior limb. Thus, this skeleton of each of the paired locomotor appendages forms a column, at whose distal end is the foot, which touches the ground. This feature emerged at the origin of the Sarcopterygii and originally was not used for body elevation, as it is possible to verify in basal groups, or relatively basal groups of sarcopterygians, represented today by the coelacanth *Latimeria* (*fig.* 2-1A) and the dipnoans *Lepidosiren*, *Protopterus* (*figs.* 2-1D, 5-10) and *Neoceratodus*.

Therefore, these structures originally had other functions. These paired locomotor appendages (even when in the form of fins) are muscular, that is, have intrinsic muscles, which do not occur in other vertebrates. These muscles are fundamental for walking but, as living basal forms do not walk, their original function must have been different. Finally, a single bone connecting paired

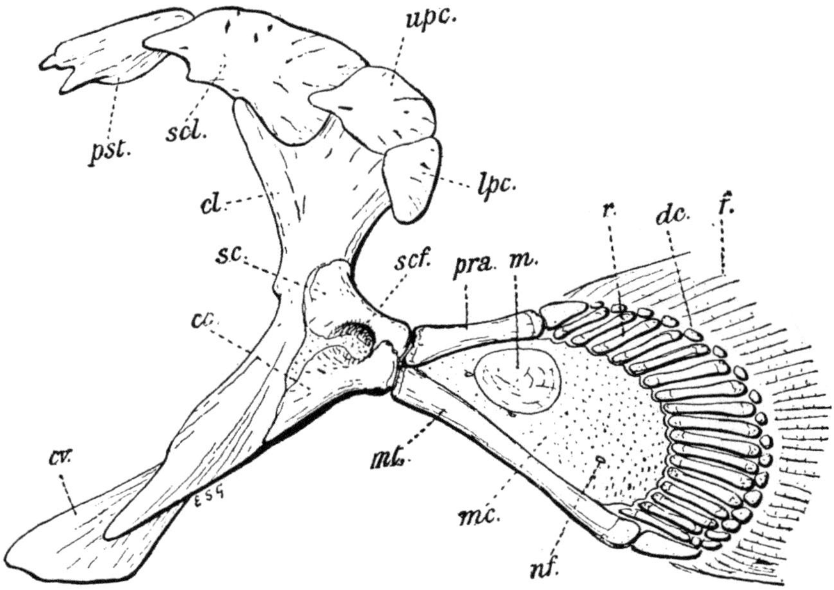

FIGURE 5-13 Inner view of skeleton of pectoral girdle of the right side of the "bichir", an African fresh-water fish (Cladistia, Actinopterygii), anterior region to the left, revealing the structure of the pectoral fin and of the pectoral girdle proper. Features are: cl, cleithrum; co, coracoid; cv, clavicle; dc, distal radial cartilage; f, dermal rays; lpc, lower postecleithrum; m, mesopterygial bony plate; mc, mesopterygial cartilage; mt, metapterygium or postaxial radial; nf, nerve foramen; pra, preaxial radial; pst, post-temporal; r, radial; sc, scapula; scf, scapular foramen; scl, supracleithrum; upc, upper postcleithrum. Connecting the post-temporal and the cranium we have the extra-scapular, not illustrated.
FROM GOODRICH (1909). IMAGE IN THE PUBLIC DOMAIN

locomotor appendages to girdles is fundamental for walking. Because there is only one skeletal element in this position in each limb, the articulations may be rounded (ball-and-socket type), and the limb can rotate nearly in any direction. On the other hand, in non-sarcopterygian gnathostomes, the bases of the paired locomotory appendages have three or more elements in this position (*fig.* 5-13), so that the articulation of these fins with their respective sustaining girdles are dorso-ventrally elongated, restricting particular movements necessary for walking. Actinopterygians usually have three skeletal elements lying next to each other in the paired fins, which unite the pectoral and pelvic girdles with their respective fins. Imagine that you have three humeri uniting your pectoral girdle to your arm. The articulation of the three bones with the

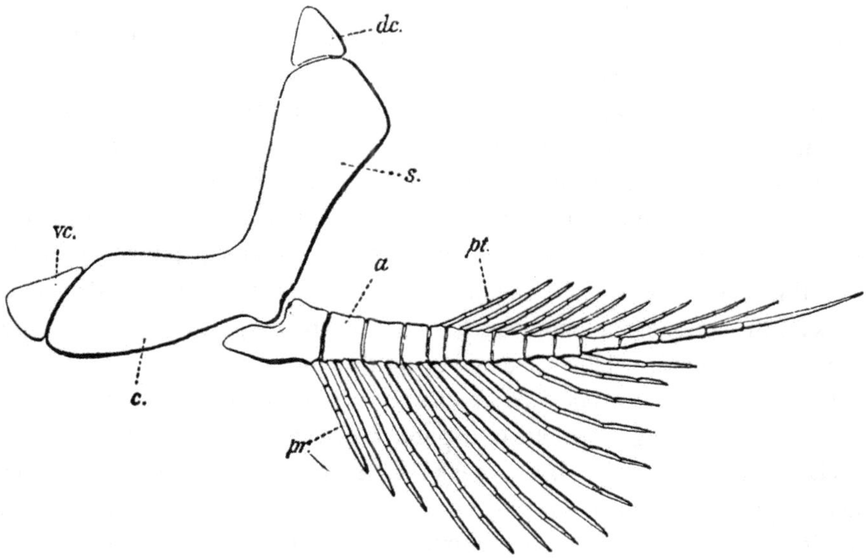

FIGURE 5-14 Lateral view of the left fin skeleton of the fossil shark †*Pleuracanthus guaudryii* (Chondrichthyes), anterior region to the lef. Features are a, segmented skeletal axis of fin; c, coracoid region; dc, dorsal cartilage; pr., pre-axial radial; pt, post-axial radial; s, scapular region; vc, ventral cartilage
FROM GOODRICH (1909). IMAGE IN THE PUBLIC DOMAIN

shoulder would become elongate, to accommodate the three humeri, and this articulation would end by losing a large part of its mobility along the axis of the articulation. With a single humerus it is possible to rotate the arm. It becomes possible, for example, in a volleyball game, to apply a "spike" movement (hitting the ball at a strong downward angle into the opponent court). But with the elongate articulation, it would only be possible to move the arm forwards and backwards, in a single plane, as do popular aquarium fish. The importance of the presence of only a single skeletal element at the basis of these locomotor limbs for walking thus becomes clear. It becomes possible to rotate the limb in all planes, moving a leg backwards, upwards and forwards, and then forwards and downwards, until it touches the soil, and then rotate in a circularly inverse direction while the body is elevated and thrust forward, ending the cycle.

As phylogenetically basal living representatives of sarcopterygians do not walk, the original function of the paired locomotor appendages must have been different. Although these structures emerged at the origin of Sarcopterygii – and are thus considered as synapomorphies of this group – they did not evolve for terrestrial walking, as argued above, once terrestriality appeared only much more recently in the cladogram (*fig.* 5-1), in derived tetrapods. Terrestriality is much more recent than the basal sarcopterygians, and even much more recent than the basal tetrapods.

The bones radius and ulna/tibia and fibula are approximately parallel, articulated respectively with carpals/tarsals, making the paired locomotor appendages (that now can also be called a limb) stronger and more suitable for walking. Yet these limbs originated in aquatic forms, as may be seen in †*Acanthostega*, and were later eventually exapted in evolution for walking. Therefore, according to Blackburn (2002), all these characters must be considered exaptations for the conquest of land by vertebrates. Several exaptations related to the humerus and femur will be discussed in another publication.

Foot: carpals, metacarpals/tarsals, metatarsals and digits (10, 111). The paired locomotor appendages present in Dipnoi, †Porolepiformes and Actinistia (= Coelacanthiformes) (*e.g.* Clack 2012: 53, *fig.* 2-14) are progressively narrower distally (toward their tips), as may be seen in *figs.* 2-1, 5-9, 5-10 and 5-11, being possibly less effective than the roundish paired fins of actinopterygians in terms of maneuverability (*figs.* 5-7 and 5-15) and fusiform sarcopterygians such as †*Eusthenopteron* (*fig.* 5-12). This is because the water foil area is progressively smaller – tapering toward the distal tip – and would benefit from a distal rounded expansion in terms of use both as a paddle and/or as a water foil to direct forward movement of the body. The distal change to a rounded "foot" was attained in water forms such as †*Acanthostega* (*fig.* 5-16), maybe by selective forces, for use underwater as foils or paddles about 395 Myr (Niedzwiedzki *et al.* 2010: 43). Several authors (*e.g.* Clack 2012, among many others) noted that †*Acanthostega* has characters of the branchial skeleton that indicate that this fossil vertebrate with feet was necessarily aquatic, because they had functional gills, which are the respiratory organs of aquatic vertebrates. We may thus interpret the toes and fingers of the feet as being structures that support the interdigital membrane, analogously to the fin rays of Actinopterygii (*fig.* 5-15) and the toes of frogs, clearly visible in the genera *Xenopus* and *Pipa* (*e.g.* Pough *et al.* 2005: 245, *fig.* 10-21f). Later, around 360 Myr, this structure was exapted for terrestriality, about 35 Myr or more after its origin. The emergence of a feature with one function, that is later coopted for another function, is enough to characterize an exaptation. This reasoning can be used to infer the exaptive nature of feet for terrestriality, following the method of Blackburn (2002; see also Laurin 2010: 88, 135).

Pelvic girdle (8, 111). Developed skeletal girdles are necessary for walking, because they are responsible for the connection of the locomotor limbs with the remaining body, permitting the transfer of the body weight to the paired locomotor appendages, including the feet, and then toward the soil, promoting

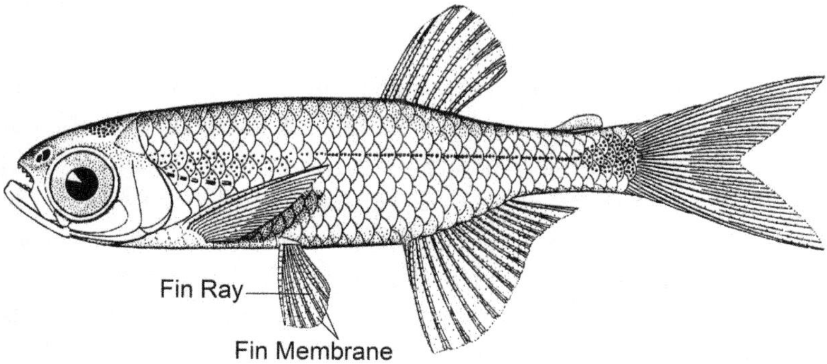

FIGURE 5-15 Lateral view of *Microbrycon minutus*, holotype, 24 mm standard length (Actinopterygii, Teleostei, Ostariophysii, Characiformes). Pelvic (also called ventral) fin with indications for fin ray and fin membranes surrounding it. Note that the membrane of the fins extends to the distal extremity of the rays of the fins, corroborating our interpretation that fingers of ancestral tetrapods functioned in support of the membranes of its paired locomotor appendages (pectoral and pelvic fins) that then could act as hydrofoils, such as in **Figure 5-16.**
MODIFIED FROM EIGENMANN (1921, PL. 89, *FIG.* 1). IMAGE IN THE PUBLIC DOMAIN

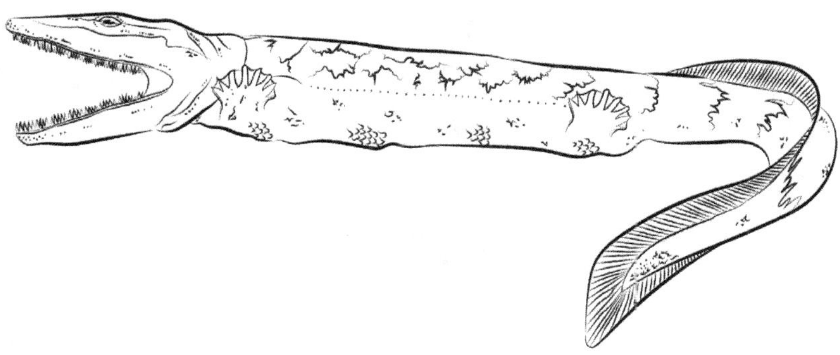

FIGURE 5-16 Artistic reconstruction of †*Acanthostega* by Alessandra Amadori, based on Triques and Christoffersen (2017: 167, *fig.* 5-16). The feet here are interpreted functionally as the fins and its fingers are considered to have a function analogous to the fin rays of actinopterygians (see **Figure 5-15**, but also **5-7 and 5-12**). Because the pectoral and pelvic fins are positioned laterally on the body, in lateral view they appear to be shorter than they actually are. The (tubular) nares, size of exposed area of eyes and undulation of caudal fin are conjectures. Scales and bone sculpture of head were omitted. Largely based on information present in Clack (2002, 2012) and Coates and Clack (no date, *ap.* Pough *et al.* 1999).

the elevation of the body away from the ground and, finally, promote walking. However, the pectoral fin girdle is already well developed in non-sarcopterygian fishes (*e.g.* Weitzman 1962 among many others; see *figs.* 1-7, 5-2, 5-7, 5-10, 5-13 and 5-14) but the pelvic girdle remained undeveloped until the level of basal tetrapodomorphs (*e.g.* Shubin *et al.* 2014: 898, *fig.* 5), as may be seen in figs. 5-2, 5-4, 5-10, 5-11 and 5-12. Recently it was shown that a developed pelvic girdle may be a synapomorphy for "†Panderichthyidae" together with more recent sarcopterygians in the cladogram, including Tetrapoda (Shubin *et al.* 2014). The developed pelvic girdle (*fig.* 5-17) is an exaptation for terrestriality, as it is already present in aquatic forms, such as †*Tiktaalik* and †*Acanthostega*, indeed in "†Panderichthyidae" and basal tetrapods. Furthermore, Clack (2012: 438) states that adaptation of the sacral region for terrestriality (being capable of bearing weight and of transmitting limb propulsion to the vertebral column) can be seen as a gradual process, with the earliest tetrapods having little in the way of a firm connection between the girdle and the vertebral column. The statement indicates the emergence of the feature as an exaptation for terrestriality. Only much later did this structure finally adapt to function in terrestrial

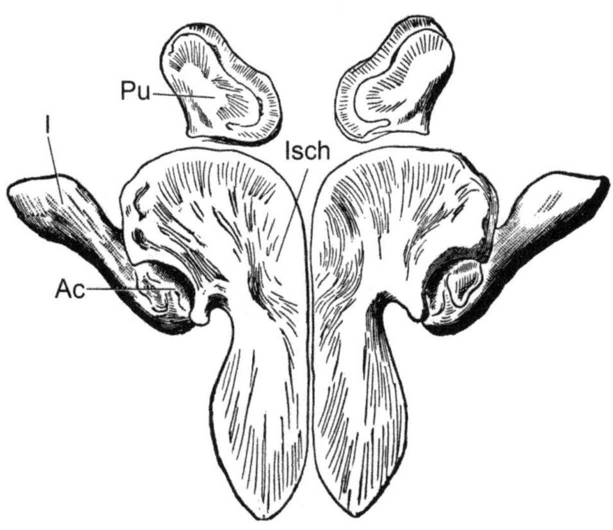

FIGURE 5-17 Ventral view of pelvic girdle (both sides illustrated) of fossil "amphibian" †*Mastodonsaurus giganteus* (Temnospondyli, Batrachomorpha, Tetrapoda), anterior region above; indicated structures are: Ac, acetabulum, which is the place in which the head of the femur inserts; I, ilium; Isch, ischium; Pu, pubis
MODIFIED FROM ZITTEL (1911). IMAGE IN THE PUBLIC DOMAIN

environments. The first tetrapods had little in the way of a firm connection between girdle and vertebral column. However, the closest known relative of tetrapods, †*Tiktaalik*, already had a large pelvic girdle (Shubin *et al.* 2014). The whole structure only became functional on land much later (several steps or nodes later in the cladogram). Therefore, according to Blackburn (2002), a developed pelvic girdle must be considered an exaptation for the conquest of land by Tetrapoda (see also Laurin 2010: 138).

Column rigidity (9, 15, 111). It has long been recognized that some bones may confer additional strength to the column, enhancing column rigidity. This may be the case of teleost inter-muscular bones, the epineurals and epipleurals (*e.g.* Weitzman 1962). Certainly this is the case of interlocking zygapophyses of powerful swimmers, such as tunas (*e.g.* Bond 1996: 73), Tetrapoda (*e.g.* Romer and Parsons 1977: 162), including †*Whatcheeria* (†Watcheeriidae) (Clack 2002: 202, *fig.* 7-6; 2012: 274, *fig.* 7-7), †*Pederpes* (†Watcheeriidae) (Clack 2012: 271, *fig.* 7-5), and of the expanded ribs of †*Tiktaalik* (Shubin *et al.* 2014: 898, *fig.* 6), †*Ichthyostega* (*e.g.* Benton 2005: 81, *fig.* 4-5d), †*Eryops* (Amphibia, Temnospondyli) (*e.g.* Benton 2005: 100, *fig.* 4-17a), and possibly many others. The interlocking of zygapophyses (as well as presumably the expanded ribs and the non-interlocking intermuscular bones of teleosts) prevents excessive rotation of vertebrae with each other (Bond 1996: 73) and is obviously useful not only for terrestrial tetrapods, but also for species with completely different life styles such as tunas. Therefore, interlocking zygapophyses (*figs.* 5-18, 5-19) may have been useful for the clearly fully aquatic colosteids (Clack 2012: 281),

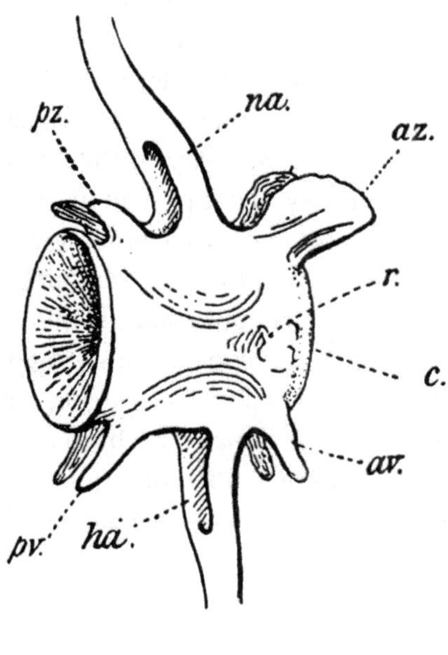

FIGURE 5-18
Caudal vertebra of tuna fish *Thynnus vulgaris* in lateral view, anterior region to the right. Structures are av, anterior ventral process; az, anterior zygapophysis or pre-zygapophysis; av, anterior ventral process; c, vertebral centrum; ha, haemal arch; na, neural arch, through which the dorsal nerve chord passes; pv, ventral posterior process; pz, posterior zygapophysis or post-zygapophysis
FROM GOODRICH (1909). IMAGE IN THE PUBLIC DOMAIN

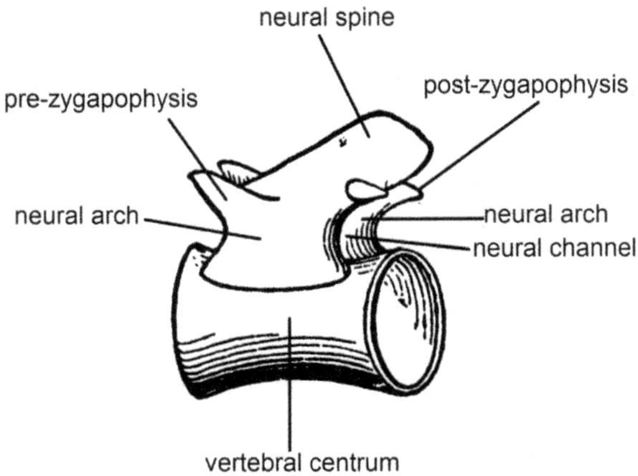

FIGURE 5-19
Lateral view of a precaudal (abdominal) schematic vertebra of †*Hyloplesion* (Tetrapoda, Neotetrapoda, Batrachomorpha, †Microsauria), showing basic vertebral features of land vertebrates. Anterior to the left
MODIFIED FROM WOODWARD (1898). IMAGE IN THE PUBLIC DOMAIN

as well as the presumably aquatic whatcheeriids, which had paddle-shaped limbs (Clack 2002: 202–203). Because limbs are not paddle-shaped in †*Pederpes* (Clack 2012: 274), a possibly non-monophyletic composition of the family is indicated (Clack 2012: 270). Thus, adaptively correlating the evolution of interlocking pre- and post-zygapophyses to the conquest of the terrestrial habitat is not possible and the same can be stated for the expanded interlocking ribs. Following the method of Blackburn (2002), it is realized that interlocking zygapophyses emerged prior to the conquest of land by Tetrapoda and must be considered an exaptation for terrestriality in Tetrapoda.

Lungs (1, III). Some authors consider the presence of lungs as a synapomorphy for Osteichthyes (*e.g.* Pough *et al.* 2005, Pough and Janis 2019: 45, *fig.* 3-5). We follow them in this regard, but the following statements of Romer (1966: 33) are noteworthy: in some specimens of †antiarchs (†Placodermi), traces of soft internal structures have been preserved; of great interest are distinct traces of a pair of large sacs connected with the floor of the pharynx. These can only be interpreted as lungs. Such structures, we believe, were present in early bony fishes of all sorts, but they are absent in sharks and have been thought to be a relatively late development in fishes. The evidence from the †antiarchs suggests, however, that they may have arisen at a much lower stage in vertebrate evolution than was previously suspected. Benton (2005: 58) also states that one specimen of †*Bothriolepis* (†Placodermi) shows possible evidence for lungs preserved inside the dorsal armor and the animal may have been able to survive in stagnant ponds by breathing air (but see a different opinion in Goujet 2011). Therefore, following Blackburn (2002), lungs must be considered an exaptation for the conquest of land by Tetrapoda.

Additional comments. According to Johansen (1970: 361), natural selection of air-breathing habits in fishes has occurred many times and in diverse ways during vertebrate evolution. The majority of extant fishes showing structural adaptations for direct use of atmospheric oxygen are tropical freshwater or estuarine forms (Johansen 1970: 361). Most air-breathing fishes are teleosts but typically the dipnoan, chondrostean, and "holostean" fishes (quotation marks are ours, as this is a paraphyletic group) are dominated by air-breathing forms (Johansen 1970: 361–362). Reports of air-breathing habits among elasmobranch fishes have not been adequately confirmed. A smaller number of air-breathing fishes live in temperate regions, and in exceptional cases, such as the bowfin, *Amia calva*, they occupy waters that are seasonally frozen (Johansen 1970: 362). Johansen (1970) recognizes three types of air-breathing organs in vertebrates: structural derivatives of the mouth and pharynx, structural adaptations of the gastrointestinal tract and the air bladder. At the time, a bladder in fishes was already generally conceded to represent an accessory respiratory organ. Therefore, we can conclude that the possibilities for respiratory adaptation, and thus for terrestriality amongst osteichthyan vertebrates, including actinopterygians, is very widespread. The probability for gaining ground in this large vertebrate group is high. However, only one group developed terrestriality, as we defined in this book. Thus, the adaptationist view of terrestriality is contradicted by these facts alone. It is not even necessary to invoke the other arguments we developed (Triques and Christoffersen 2009, 2015, 2017, 2018) and that are presented here. More exceptionally, fishes like *Periophthalmus* have developed air-breathing habits, in support of behavioral acts such as feeding and escape reactions (Johansen 1970: 362).

Johansen (1970: 363) considers that structural adaptations for air breathing in fishes are extremely diverse. The extent of modifications from structures typical of an aquatic breather often reflects the relative role of air breathing in overall gas exchange. Some species rely predominantly on water breathing with gills, and only supplement gas exchange with air breathing when adverse respiratory qualities in the water make aquatic breathing insufficient or too costly for extraction of the required oxygen. Others are obligate air breathers, and succumb if denied access to air for a short time (Johansen 1970: 363). We consider that the "extreme" diversity in structural adaptations for breathing air in osteichthyans, as evaluated by Johansen (1970), is currently, and was through geologic time, an "extremely" rich substrate for selective pressures to work on. Had such pressures been responsible for gaining ground in vertebrates, the terrestrial vertebrates should be polyphyletic, because respiratory traits for air breathing were already present since the origins of osteichthyans. More significantly, air breathing devices seem to have appeared independently from each other several times (except for the lungs, which appeared only once), and with somehow different natures, as noted by Johansen (1970: 363–375).

In dipnoan fishes the air bladder (as Johansen defines the organ) has been retained and further evolved the respiratory function it allegedly possessed in crossopterygian fishes (a paraphyletic group of sarcopterygians, a term we do not use in this book). Amphibians and land vertebrates later arose from this lineage (Johansen 1970: 371). Members of the primitive orders Chondrostei (*Polypterus*) and Holostei (*Amia* and *Lepisosteus*) have also retained a respiratory function of the air bladder, but fall short of the refinement in structural adjustment for air breathing seen in the dipnoans (Johansen 1970: 372).

Only the lungfishes, and of these *Protopterus* (the African living genus) and *Lepidosiren* (the South American living genus), in particular, have the air bladder evolved structurally to resemble a lung, as it exists in lower vertebrates (Johansen 1970: 372). We consider that anagenesis, that is, evolution within a phyletic lineage, may produce marked differences between different evolutionary branches. For instance, Aves and Crocodilia are markedly different taxonomic groups, in morphological, ecological, physiological and behavioral terms (and maybe others as well), but they are yet archosaurian amniotes, together with dinosaurs and pterosaurs. Therefore, lung differences between actinopterygians and sarcopterygian osteichthyans, in lung morphology and functionality, should be expected, but this does not mean that these lungs should be interpreted as analogous instead of homologous structures. Other researchers, since before Johansen (1970), continue considering a single origin for the lung, that later became an air bladder in teleost actinopterygians. The incapacity of perceiving homology relations between features or states of a feature would only lead to a gradistic point of view that ultimately would be a return to gradistic (and creationist-like) classifications of living beings, with outdated interpretations of evolution. An adaptationist view of evolution could be sustained by the result of this reasoning. The capacity of perceiving homologies and the homologous nature of all known lungs favors other possibilities for the understanding of evolution and the evolutionary process. In this case specifically, it allows the visualization that the evolutionary process includes exaptations, and that the lungs are exaptations for terrestriality.

Johansen (1970: 372–373, *fig.* 3) describes and illustrates the impressively developed lungs of lungfishes. Internal septa, ridges, and pillars divide the air space into smaller compartments opening into a median cavity. Further subdivision of these compartments by progressively smaller reticulated septa terminates in alveoli-like pockets, richly covered with blood vessels (Johansen 1970: 372). *Neoceratodus* (the Australian living genus) has one lung, while *Protopterus* and *Lepidosiren* have two lungs, fused in the anterior region, where the pneumatic duct and the pneumatic sphincter connect the lungs with the esophagus. The short airway duct, which in no way resembles the air distribution system of the trachea and bronchi of terrestrial vertebrates, is richly supplied with smooth muscle, as is the lung parenchyma itself (Johansen 1970: 372). The smooth musculature is important for the mechanics of breathing and may exert additional influence in distributing the air inside the lung (Grigg 1965, Johansen and Reite 1967 *ap.* Johansen 1970: 372). Electron microscopy has revealed that the fine structure of the lung in *Protopterus* does not differ basically from that in higher vertebrates (DeGoodt *et al.* 1960, Schulz 1960 *ap.* Johansen 1970: 372). The respiratory epithelium adjoins the basement membrane, which in turn is separated from the endothelial basement membrane by a thin layer of collagen fibrils (Klika and Leleck 1967 *ap.* Johansen 970: 372). The *Protopterus* lung is much more similar to the amphibian lung than to the lung of *Polypterus*, which has a smooth inner surface, that is much less vascular than in the lungfishes (Johansen 1970: 372).

Only fishes employing the air bladder for aerial gas exchange have specialized vessels developed as connections between the primary branchial circulation and the gas exchange organ (Johansen 1970: 375). These vessels take origin from the sixth aortic arch and are homologous to the pulmonary arteries of higher vertebrates. In addition to the dipnoans, *Polypterus* and *Amia* possess these vessels (Johansen 1970: 375). We take advantage of these two statements by Johansen (1970) to indicate that even vascular details show homologies between the "respiratory organs" of actinopterygians and sarcopterygians, showing their common origin, that is, they are homologous. These organs thus deserve to receive the same name, which thus indicates common origin and a relation of homology between them: the lung.

Subsequently, Johansen (1970: 375) indicates differences between sarcopterygians and actinopterygians, which we interpret as being autapomorphies of different phyletic lineages, as follows. The dipnoans are unique among fishes by also having a specialized efferent circulation from the air bladder, connected directly with the left side of the heart. *Gymnarchus niloticus*, a teleost using the air bladder for accessory respiration, has an arterial supply to the bladder from the dorsal aorta, but a venous drainage connecting directly with the atrium.

Again, we emphasize that researchers working on any evolutionary interpretation need to understand and differentiate between homology and differences that do not indicate homology but also do not contradict notorious homologies. By making a simple analogy, the fact that Aves have feathers indicates an apomorphy for Aves or Aves together with a subset of dinosaurs (as has been suggested multiple times in the recent literature), but this apomorphy does not indicate that Aves or Aves plus a part of or the totality of the dinosaurs do not have phyletic relationships with crocodilians, forming Archosauria or Archosauromorpha. That is, an apomorphy for a less comprehensive group does not necessarily invalidate the apomorphies for larger groups. In a similar vein, the "single" characteristics of dipnoans do not invalidate the homology found between specialized vessels developed as connections between the primary branchial circulation and the gaseous exchange organ (lung), which originate from the sixth aortic arch, and which are homologous to the pulmonary arteries of the "higher" vertebrates, as indicated in the previous paragraph.

No septation of the heart is evident [in actinopterygians and fish-like sarcopterygians] (Johansen 1970: 375). However, the heart in dipnoans is in part structurally adapted for accommodation of two bloodstreams, by showing partial atrial and ventricular septa, structures which clearly forecast a further development of the heart and circulation toward conditions present in terrestrial vertebrates (Johansen and Hanson 1968, Johansen and Hol 1968 *ap.* Johansen 1970: 375). It is clear that this is one more exaptation for terrestriality, which we could present as a distinct character in this book.

Choanae (5, III). These are the paired internal openings on the "roof of the mouth", or "palate", for the entering of air to the lungs (*figs.* 5-5, 5-6, and 5-20). The air enters by the external nares (one to each side of the snout), follows within a short canal to the interior of the mouth, where it may enter because of the inner openings, the choanae. According to Janvier (2004) they are homologous to the posterior nostrils present on each side of the snout in coelacanths and teleosts, which have independently migrated downward and finally reached the inner side of mouth in Dipnoi and Tetrapodomorpha. They appear in the cladogram (*fig.* 5-1) much before tetrapods attained terrestriality, and thus must be considered exaptations for the conquest of land by Tetrapoda, according to the method of Blackburn (2002). Additionally, Clack (2012: 90–91) states that the external nostril of †*Acanthostega* is small enough to suggest that it was not used for the intake of air in breathing, being used as in basal fishlike species (that is, as an olfactory organ), reinforcing the proposition that this feature was an exaptation for terrestriality. Clack (2012: 388) considers that the external nostril (on each side of the snout) may have been enlarged rapidly in the Carboniferous forms. Therefore, the use for respiratory purposes of the internal nostril (the choana, obviously connected to the external one) should be interpreted as an exaptation for terrestriality, instead of an adaptation.

Alternating movement of paired locomotor appendages (2, III). The emergence of lungs in Osteichthyes and its derivative swimming bladder in teleosts conferred these animals with buoyancy. Buoyancy in the water column gives the animal the possibility to maneuver delicately. Delicate maneuvers, however, are best made with alternating motions of the paired locomotor appendages. Thus, we link the origin of the alternating movements of these

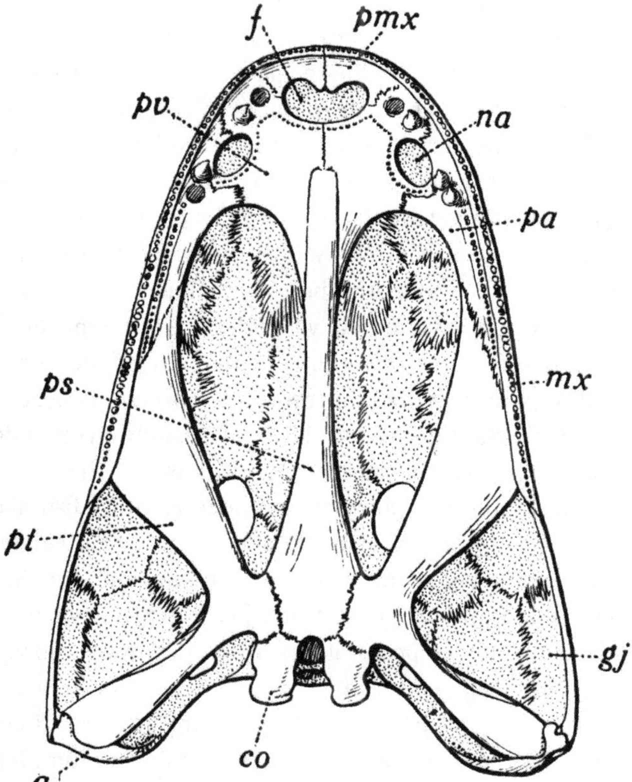

FIGURE 5-20 Ventral view of roof of mouth, the palatum of the fossil amphibian †*Cyclotosaurus robustus* (Temnospondyli, Batrachomorpha, Tetrapoda). Features are co, condyle on exoccipital; f, foramen for teeth of lower jaw; gj, quadratojugal; mx, maxillary; pa, palatine; na, internal nare, also known as choana; pms, premaxillary; ps, parasphenoid; pt, pterygoid; pv, prevomer; q, quadrate
FROM FRAAS (1889 AP. GOODRICH 1930). IMAGE IN THE PUBLIC DOMAIN

appendages to the appearance of the lung in osteichthyans. The alternating motion of the paired locomotor appendages is then interpreted by us as a by-product of the appearance of the lungs. Thus, the need for alternate movement of paired fins may be explained. Considering the lung-less or bladder-less chimaeras and manta rays as good out-groups to ordinate the character direction, the plesiomorphic condition of the movement of the paired locomotor appendages would be the coordinated movements of paired fins up and down (*e.g.* Lagler *et al.* 1962: 191, *fig.* 6-2.d), both in ordinary movements or in foraging. Therefore, the observed behavior of alternating paired fins in non-walking sarcopterygians is herein considered only a by-product of the emergence of the lungs/gas bladder. This feature occurs also in actinopterygians, as may be seen clearly in freshwater angelfish (*Pterophyllum*), sold in aquarium shops. That is, a buoyancy organ cancels body density relative to water. Therefore, the necessity arises for equilibrium by moving slowly. If the fins moved disorderly, the body would lose stability in the water. Another way to illustrate the situation: A man in a canoe will use paddles alternately in order to go straight ahead evenly. If one paddle is used more than the other, the direction of the movement turns. If both paddles are used at the same time, the movement ahead is of the stop-and-go kind, not constantly ahead. Paddling alternately on one side and the other is the form of propulsion that continuously moves the canoe forward. The same reasoning applies to aquatic vertebrates. A fish or fish-like animal using paired fins to move slowly becomes necessary (using the caudal fin for strong thrust and therefore fast swimming usually implies another kind of use of the paired locomotor appendages – as water foils). In fact, even actinopterygians alternate the position of pectoral fins, when they move slowly. As expected, occasionally any living species with paired locomotor appendages can use their paired locomotor appendages together in the same direction and at the same time. Like humans, when breast-swimming, or jumping in a basketball game. Mammals, when galloping or simply running, as in kangaroos, provide another example. But we consider the original movement of paired locomotor appendages in osteichthyans as being of the alternate pattern during the slow movement of the animal, as explained above. Therefore, the trackways of *manus* and *pes* prints found in Poland by Niedzwiedzki *et al.* (2010) were predictably of the alternating pattern, as encountered by these authors. They move in the very same way as teleost ogcocephalid batfishes (actinopterygian fishes, *fig.* 3-1) do: the paired fins move independently in a diagonal sequence of steps, according to Rade (2013). We do not consider this type of movement as an adaptation for walking, correlated to the conquest of land. Thus, the "dog-like" movements of paired fins of *Latimeria*, *Lepidosiren* (Dipnoi) (*e.g.* Janvier 1996: 74), and the trackways from Poland, are here considered as homologous to the alternating movements of actinopterygians in slow maneuvering. These alternating movements may be an osteichthyian feature, or even a more ancient behavior. Therefore, the alternating movement of paired locomotor appendages is here considered an exaptation for the conquest of land, following the method initially proposed by Blackburn (2002). Regarding this theme, see also Fricke *et al.* (1987).

Alternatively, the alternating movement of paired appendages in terrestrial vertebrates may be considered as an exaptation from the undulatory movement of "fishes", including sharks (*e.g.* Dean 1895: 1–2, *fig.* 1-2) and other fish-like animals. The application of the method of Blackburn (2002) renders the same result as the previous proposition.

Additional comments. Baerends (1971) conducts an ethological analysis of fish behavior. He considers that ethology is the study of all aspects of behavior using biological methods. These aspects can be considered as belonging to five categories, according to him: the description, the causation, the ontogeny, the function or survival value, and the evolution of the group. Problems of causation can often be better approached against a background of knowledge about the function of the behavioral patterns, and can sometimes benefit from ideas on their evolution (Baerends 1971: 279–280). When concentrating on one aspect, it is advantageous to remain open-minded for other possibilities, provided one always carefully avoids mixing up problems and answers of different categories (Baerends 1971: 280). Our analysis of the alternating movement of paired appendages of osteichthyans and, more emphatically, of sarcopterygian vertebrates, follows the considerations of Baerends (1971). We described the behavior and suggested a causation factor: the emergence of an internal air vesicle organ, the lung or its derivative, the gas bladder, in a more recent hierarchical level, the actinopterygian teleosts. This causal proposition is new in the literature and disagrees with previous authors. We also suggested a faction that is different from proposals in the current literature. We then interpreted the complete set of aspects with a cladistic background, and analyzed the possibilities according to the method of Blackburn (2002), maintaining the focus of our book: vertebrate terrestriality. We agree with all proposals made by Baerends (1971). We consider that the integrative analysis he proposes to follow among each component of behavior, benefits a new, divergent point of view, or may clarify a preexisting perspective. To Baerends (1971: 280), behavior is usually a very complicated phenomenon, through which the animal is capable of adjusting its various functions to a constant or changing environment. What we have found conforms exactly to this point of view: the alternating movement of paired appendages adjusts the animal movement in slow fashion to the presence of an internal buoyancy organ, even in the case of lungs being originally a respiratory instead of a buoyancy organ. This occurs in a "constant" environment, that is, in slow moving water, or in a non-moving water body. By isolating smaller components, the researcher can reduce the number of intervening variables, and thus increase the exactness of his work (Baerends 1971: 280). However, one cannot understand the functioning of an entire machinery only from a detailed study of its parts (Baerends 1971: 280). We may better understand the exaptive nature of the alternating movements of paired appendages when the interactively interpreted locomotory appendages are correlated with the buoyancy or buoyance affecting the organ. Baerends (1971) concludes that, in order to obtain an insight on how these parts work together, one must also start from the entity behavior and analyze it downward to the level of simple parts. This is what causal ethological analysis is trying to do: to split up a behavioral complex, as it appears to the observer, into units – often of different levels of integration – with circumscribed tasks, and sufficiently restricted to be an acceptable challenge for attack by the researcher [the author talks to physiologists, but we include ourselves as researchers he talks to] with his/her methods (Baerends 1971: 280). Evolution is modified ontogeny (De Beer 1940 *ap.* Baerends 1971: 348), and ontogeny is a chain of causal processes. Consequently, knowledge of the causation of behavior may be expected to help understanding how the different behavioral patterns have evolved (Baerends, 1958 *ap.* Baerends 1971: 348). Our analysis of alternating movements of paired appendages fits clearly into the statement that the knowledge of causation (the buoyancy organ) of behavior (alternating movements) helps us to understand this last factor of evolution, and its exaptive nature regarding vertebrate terrestriality.

We used the chimaeriform and rajiform chondrichthyans as possible or practical outgroups to polarize the alternating movement of paired locomotor appendages in slow moving locomotion, as an effect of the emergence of the lungs, and consequently of buoyancy. As †placoderms and †acanthodians are extinct animals, we do not have other possibilities for this kind of comparison among the gnathostomes and jaw-less vertebrates with paired fins, which are also extinct: †Thelodontiformes (or †Coelolepididiformes), †Anaspida, †Cephalaspidiformes (or †Osteostraci) and †Pituriaspidiformes (†Pituriaspida) (Nelson 2006). In order to complement our initial presentation on their mode of locomotion, we will briefly present the rajiform mode of locomotion, as proposed by Lindsey (1978: 35–36, *fig.* 6). The primary locomotor organs in most rays, order Rajiformes, are the greatly enlarged pectorals, which form a wide lateral expansion of the body. Waves of undulation in a vertical plane are passed backward along the mobile fin margins ("rajiform mode"). In the majority of rays, the only organs other than the pectoral fins which may contribute at all to forward locomotion are the pelvic fins. These lie at the posterior end of the disc created by the snout and pectoral margins, and may be used to kick back as the ray is taking off from the bottom. Two families, the eagle rays, Myliobatidae, and the mantas, Mobbulidae, have the pectorals even more expanded than in the skates and rays, so that their width is greater than the length of the main part of the body. These animals are secondarily adapted for a free-swimming, rather than to a sedentary existence (bottom dwelling). Their tremendously mobile pectorals produce a graceful rajiform swimming motion (Lindsey 1978: 35, *fig.* 6), analogous to the flight of birds. Longitudinal sections through the body show it to be constituted by a series of hydrofoils. The base of the pectorals assumes a sharp angle of attack during the upstroke of the wing tips (Klausewits 1964 *ap.* Lindsey 1978: 36), providing a lift to overcome the excess of weight over buoyancy. Chimaeras (Holocephali) are considered to have two types d pectoral locomotion (Breder 1926 *ap.* Lindsey 1978). They may row through the water or they may hold the pectorals out at right angles to the body, and undulate them in the rajiform mode. The labriform mode of locomotion regards the presence of narrow pectorals with nearly vertical bases, and the pectorals are flapped rapidly and synchronously, often being fanlike and rounded. They may also undulate (Lindsey 1978: 31). Their pectorals are not as wide-based as in Rajiformes, and the pectoral locomotion described by Breder is perhaps closer to the diodontiform mode (with a posterior anguilliform undulation of the body). Therefore, in chimaeras and rays, the synchronic up-and-down movement of pectoral fins is the primary mode for obtaining propulsion, in opposition to osteichthyans, as we argued above.

Functional neck (II, 12–13, 16, III). The neck is advantageous for land vertebrates, as it permits the head to rotate, making it easier to feed and to seek around without affecting locomotion (Hildebrand 1982: 57). The development of a functional neck apparently occurred through several distinct steps. The loss of opercular bones permitted an up and down bending of the head more freely. Such an advantage is unknown for aquatic vertebrates, which are benefitted by suction (and/or other kinds of uses of the prehensile jaws [*e.g.* Lauder and Liem 1983: 158–159]), alternative to *sensu stricto* biting. It is not yet clear if there was in fact any advantage in the loss of these opercular bones. Their loss alone is not sufficient for the appearance of a functional neck, as we will try to make clear below. Yet their loss is a necessary condition for the development of a functional neck. A disconnection of the pectoral girdle relative to the head is necessary to free the head from the body. It is not easy to discuss favorable terms of evolutionary advantages for this condition, as a functional neck was not yet developed at this moment in the evolution of the Tetrapoda. The notochord (a firm but flexible chord – *fig.* 5-21, NCH), which in most extant vertebrates is being substituted during development by the vertebral bone centers – *figs.* 1-5C′ and 1-7 VC) was inserted inside the skull, preventing a free movement of the head. Finally, but not less important, was the withdrawal of the notochord from inside the skull, again to free the head from the body. According to Clack (2012: 207) “the notochord is bendable, but springs naturally back to a straight line unless muscular effort is exerted. The earliest tetrapods probably had little control over the attitude of the head and probably had to expend considerable energy to move it sideways or up and down". Therefore, to make the neck a functional feature in the way land tetrapods use it, all of these conditions should be present at the same time, but the cladogram shows that they appeared successively and independently in evolution (in any case, the withdrawal of the notochord from the skull was a much later occurrence, as can be seen in *fig.* 5-1). According to the cladogram presented herein, only the withdrawal of the notochord from the skull occurred together with a functional neck. Selective forces can only be suggested at this moment, from the perspective of the method of Blackburn (2002) for gaining ground in vertebrates. In case a more basal position is preferred for the †baphetids (*e.g.* Laurin 2010: 162, *fig.* 7-1), the reasoning continues to be the same. Indeed, in this case, a selectionist explanation is reinforced, as the same feature would be emerging independently twice. It is worth noting that, in †*Tiktaalik*, the opercular series of bones appear to be completely missing, and the same may be the case for the extrascapular series (Clack 2012: 196). That is, this loss of several unrelated bones may be a synapomorphy for the Tetrapoda plus some more basal taxa that we placed within "†Panderichthyidae", but which do not appear in *fig.* 5-1. The hypothesis that these losses of extrascapular and opercle bones represent exaptations for terrestriality does not change for these characters (see also Laurin 2010: 138).

Depressiform head and body (6, 1). A body form flattened from top to bottom is usually called "depressed" (*e.g.* Lagler *et al.* 1962: 53) and, less often, "depressiform" (Bond 1996: 17). Such are the cases for skates and their relatives, angel sharks, toad fish, angler fishes, batfishes (*fig.* 3-1), as well as manta rays, for example. Such a body shape suits the fish for life on the bottom, for living under the water surface, and for flight-like swimming above the bottom. "†Panderichthyidae" (currently considered a paraphyletic assemblage of taxa progressively closer to Tetrapoda) and basal tetrapods (*fig.* 5-16) are depressiform, with dorsal eyes, and likely lived on the shore or edge of lakes and/or rivers, avoiding deep areas because of the need to breath air into their lungs. In the cladogram it is possible to regard this body form as selected for life in shallow waters, probably being an adaptation instead of an exaptation for this kind of habitat, following Blackburn's (2002) method. Thereafter, the depressiform head and body shape may not be considered adaptations for life on land, as they are already considered to be possible adaptations for shallow waters through the application of the method of Blackburn (2002). Further, the depressiform shape of the skull can hardly be correlated with selective pressures for gaining ground, because amongst the taxa known from even the earliest part of the Carboniferous (such as †*Pederpes*), we find skulls deeper

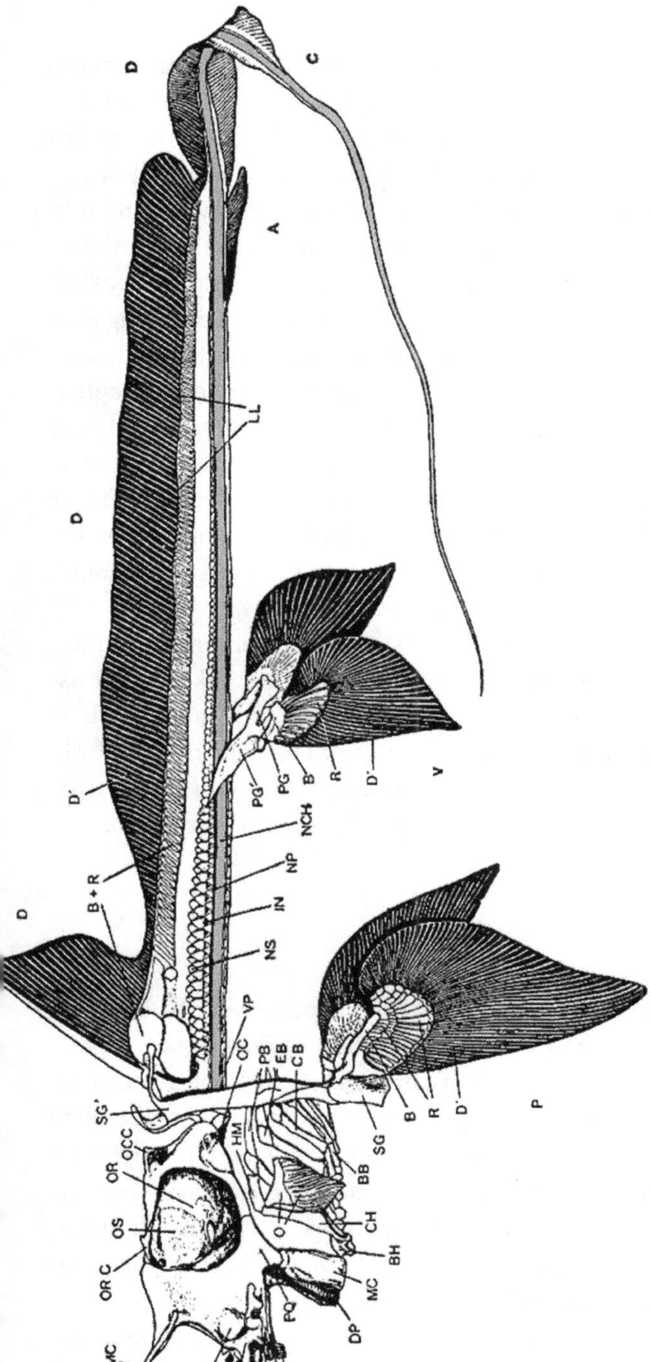

FIGURE 5-21 Skeleton of chimera (*Chimera monstrosa*, Chondrichthyes) in lateral view. A, anal fin; B, basal skeletal elements of fin; BB, basibranchial; BH, basi-hial; C, caudal fin; CB, ceratobranchial; CH, ceratohial; D, dorsal fin; D′, ceratotrichia, which are fin rays composed by proteic tissue; DP, dental plate; EB, epibranchial; HM, hyomandibular; NCH, notochord (indicated in light grey), which comes from inside of skull (not shown) and prolongs to tail tip; IN, interdorsal element of vertebra; LC, labio-nasal cartilage; LL, longitudinal ligaments; MC, Meckel's cartilage; NC, nasal capsule; NS, neural spine; NP, basidorsal element of vertebra (?); O, opercular cartilage and related opercle cartilaginous rays; OC, otic capsule; OCC, occipital crest; OR, orbit; ORC, orbital crest; OS, interorbital septum; P, pectoral fin; PB, pharyngobranchial; PG, pelvic girdle; PG′, iliac process; PQ, palate-quadrate; R, radial; SG, scapular or pectoral girdle; SG′, scapular process; UMC, rostral; V, ventral or pelvic fin; VP, fused vertebrae

MODIFIED FROM DEAN (1895). IMAGE IN THE PUBLIC DOMAIN

than broad, or conspicuously deep in †*Crassigyrinus* (†Crassigyrinidae) (Clack 2012: 388).

Loss of anal and dorsal fins (7, I, III). Dorsal (sometimes there are two fins in sequence) and anal fins (*figs*. 5-2, 5-3, 5-4, 5-7, 5-8, 5-9, and 5-10) are median appendages usually regarded as providing horizontal direction during fast movements in the water. For a bottom living species, the absence of one or both of these fins may not be too disadvantageous if the life style is ambush-like, and especially if the body shape is depressiform, in which the movement is slower and covers shorter distances. Supposedly, species living in very shallow waters could benefit from the absence of these fins by reducing drag from both the surface and the bottom of the water. This may have been the case of †panderichthyids and basal tetrapods (*fig*. 5-16), possibly being adaptive, rather than exaptive, for living on the bottom, after the use of Blackburn's (2002) method. On the other hand, the absence of these median fins should be considered exaptations for terrestriality, following this same method. These exaptations have the effect of diminishing water loss by evaporation on dry land because, in this environment, flaps would permit a better evaporative effect through the thin air.

Ribs developed and ventrally directed (14, III). Because Carboniferous lungfishes possess long and curved ribs, almost enclosing the trunk, and yet use buccal pumping for air intake, it appears that the ribs were more likely to have been involved in locomotion (Clack 2012: 429). These characters of ribs appeared in the cladogram much before terrestriality. Blackburn's (2002) procedure indicates the exaptive nature of these novelties for living on land.

3 Life Styles

The "functions" of features (or "effects" of exaptations) seem to be better referred to as "life styles", at least in some cases, in the following text. That is, life styles represent the way the animals live or use their anatomical attributes or features to deal with their environment. The life style is assumed to exist as a function (or effect) of particular features. Thus, in *fig*. 5-1, life styles are indicated with Roman numerals (I – III) in opposition to features, which are enumerated in Arabic numerals (1 through 16). The evaluation of "life styles", as representing Functions and/or Effects of structures/behaviors, using the method of Blackburn (2002), are indicated in the cladogram.

Life in shallow waters (I, a Function of 6–7). Apparently, several tetrapodomorph representatives underwent a trend toward a role as ambush predators in shallow waters (Clack 2012: 84–85), maybe correlated to the actinopterygian prey diversification through the Devonian. †*Elpistostege*, †*Tiktaalik* and †*Panderichthys* belong to the family †"Panderichthydae", and come from estuarine localities (Daeschler *et al.* 2006); †*Panderichthys*, †*Obruchevichthys* (†Elginerpetonidae: †Ichthyostegalia) and †*Ventastega* (†Ichthyostegalia) are interpreted as living in marginal localities at the edge of a large marine basin (*e.g.* Luksevics and Zupins 2004). Clack (2012: 185) considers that many basal tetrapods, such as †*Tulerpeton* (†Tulerpetonidae: †Ichthyostegalia), †*Ventastega*, †*Hynerpeton* (†Ichthyostegalidae: †Ichthyostegalia), †*Densignathus* (†Ichthyostegalia), †*Metaxygnathus* (†Ichthyostegalia), †*Sinostega* (†Ichthyostegalia), and †*Jakubsonia* (†Ichthyostegalia) lived at the water margins of lakes, deltas or coasts. Depressiform body shape and loss of the dorsal and anal fins emerged together with this life style, and may be considered as possible adaptations for life in shallow water. Following the method of Blackburn (2002), this life style may be a function of characters 6 (head and body depressiform with dorsal eyes – *fig.* 5-16) and 7 (loss of unpaired dorsal and anal fins – *fig.* 5-16).

Terrestriality (III, an Effect of 1 – 5, 7 – 15; a Function of 11 and 16). Walking with paired limbs, breathing air and feeding on land are herein summarized as "terrestriality", to differentiate this life style from crawling and foraging on land. We do not consider crawling or jumping to be true walking, and certainly has no relationship to the breathing of atmospheric air seen in some teleost fishes such as *Periophthalmus* (Actinopterygii). Armored catfishes of the family Callichthyidae are teleosts that crawl on land and diffuse air through the digestive tube (vascular hindguts), not having lungs (Nelson *et al.* 2016: 215). Walking (in opposition to crawling), in terrestrial vertebrates is accomplished with the use of paired columnar appendages, which elevate the body above the ground. These limbs are popularly known as legs. These have intrinsic musculature (character 4 in *fig.* 5-1) and an internal skeleton organized with a single skeletal element connecting to the pectoral or pelvic girdle (character 3 in *fig.* 5-1). It is followed by two parallel elements (character 11 in *fig.* 5-1), after which comes the foot (character 10 in *fig.* 5-1). The limbs (which are modified paired locomotor appendages) are capable of alternating movements, jointly producing walking movements (character 2 in *fig.* 5-1). For this, it is necessary for these limbs to have intrinsic muscles (character 4 in *fig.* 5-1), as well as a single skeletal element at its base, in order to connect with the respective girdle (character 3 in *fig.* 5-1). Thus, we should not consider only the feet and its toes (as some authors do), although they are essential to step onto soil. It is necessary to stress the necessity of columnar appendages (the limbs, as described and defined in Romer and Parsons 1977: 195–197). Many researchers apparently neglect the necessity of feet being connected with the remaining leg, forming a joint columnar appendage. The step on land is only possible with the use of the whole columnar limb (leg). It would be impossible to have functional feet, for walking on land, without legs. This differentiates walking from crawling (possible for non-sarcopterygian vertebrates). It is noteworthy that columnar (strut-like) paired walking appendages could emerge in Sarcopterygii by means of a single proximal element (character 3 in *fig.* 5-1). Its emergence in the taxon is relatively ancient, and has a sequential position (from proximal to distal in the limb) concerning the more distal radius and ulna/tibia and fibula (Romer and Parsons 1977: 195–196), which in turn are parallel to each other (character 11 in *fig.* 5-1). Furthermore, the columnar paired appendage could not be used for walking if its connection with girdles were not somehow narrow toward the body and rounded at the extremitiy that touch the girdle, that is, if they were not composed by a single skeletal element (character 3 in *fig.* 5-1). If there were more than a single proximal skeletal element on each of the paired walking appendages, and if distally there were no structure similar to a foot (character 10 in *fig.* 5-1), walking would not be possible. If the connection with the girdle (and consequently, with the body) were elongated instead of rounded, movement would be limited to the extent that walking could hardly be achieved. The lobe-fin fishes (sarcopterygians), by contrast, attach their paired fins by a single radial – homologue to the femur and humerus – to the girdles. These form the main axis of the fin by stringing the rest of the radials in a chain growing outward from the body and having a narrow, rotative base that allows the limb to rotate as well as to be raised and lowered (Clack 2002: 17). It is enough for the reader to look at his/her own legs and arms and then check how they move, to understand what was said here. Clearly, this character complex (muscular columnar appendage with a single proximal skeletal element and two subsequent elements) arose a single time, around 416 Myr, in the beginning of Devonian (Zhu and Yu 2009). Therefore, walking appendages arose prior to all the Devonian geological alterations about 56 Myr, before the conquest of land, if we consider the boundary between the Devonian and the Carboniferous as coinciding with the beginning of terrestriality. This represents too much time for microevolution to have been responsible for this accomplishment. Further, we must consider that all these characters that compose the limbs were already present, before the imagined selective pressures of the Devonian had appeared.

The alternate movements of the paired locomotor limbs (character 2 in *fig.* 5-1) appeared before the Devonian, several levels before terrestriality in the cladogram. We interpret alternate limb movements as an exaptation for terrestriality, in the same way as lungs (character 1 in *fig.* 5-1). A depressiform body with dorsal eyes (character 1 in *fig.* 5-1), which the method of Blackburn (2002) correlates with a possible process of adaptation for a life style in shallow water (life style I, in *fig.* 5-1), are evaluated as exaptations for terrestriality, by the same method of Blackburn (2002).

Although air breathing could occur without internal nares (choanae, character 5 in *fig.* 5-1), this feature diminished moisture loss to the environment to an extent that permitted the free conquest of land and full terrestriality. However, choanae emerged at about the time of origin of the Rhipidistia, not at the time of the origin of terrestriality. Zygapophyses (characters 9 and 15, in *fig.* 5-1) are of fundamental importance to avoid the vertebral column from collapsing during walking on dry land, but emerged prior to terrestriality. Zygapophyses function correlated to the developed skeletal girdles (character 8 in *fig.* 5-1), also appearing several levels before terrestriality in the cladogram. They are considered to be exaptations for the conquest of land. The loss of dorsal and anal fins (character 7 in *fig.* 5-1) is advantageous for a land vertebrate, because these structures represent widened surfaces, in which loss of moistness would harm the animal. Unwanted water loss, as well as possible damage could occur during walks through dry woods. However, these losses occurred before the emergence of feet, and much earlier than terrestriality. Therefore, these losses can only be considered exaptations for land, according to Blackburn's method. This is also logical, because it is illogical to consider any features that emerged before feet as an adaptation for terrestiality, in the cladogram.

Without feet, or feet-like features, walking on land is not possible for vertebrates, which have an endoskeleton instead of the exoskeleton of arthropod legs. In vertebrates, the existing alternative to the feet (to maneuver in water) are the fin rays, and these rays would not be capable to support weight and walk on solid ground, in animals from a certain size and weight. Therefore, any feature used for terrestrial walking is useless on land without the presence of feet. Thus, it is not possible for a feature (whatever it may be) to be an adaptation for land if feet are not already present and ready for walking. As a consequence, it becomes illogical to think of these features as adaptations for terrestriality in vertebrates. Following this argument, features 1 to 9 in the cladogram of *fig.* 5-1 cannot be viewed as adaptations for terrestriality because they emerged before feet. They could never be used on land before feet had evolved. Therefore, they are not adaptations for land, in any sense.

Independent of the results of the method of Blackburn (2002), these features, in order to be considered as adaptations for terrestrialily, would have to have emerged in evolution at the same point in the cladogram as feet. All of these features, including feet, should have emerged as "half features" for terrestriality and then progressively have been adapted for terrestriality. "Half of a foot" should be semi-circular, instead of circular, in dorsal or ventral profile view of each foot, but this is not known in the fossil record. "Half muscular" paired locomotor appendages should show progressive "entrance" of muscles inside each appendage, through geologic time and in the fossil record. And so on, feature by feature. However, this is not shown by the fossil record. Counting steps from character 10 (the feet) in the simplified cladogram of *fig.* 5-1, downward to character number 1, it is possible to see up to seven steps. Therefore, these features did not evolve together with feet, and cannot logically be considered adaptations for terrestriality. This is a completely new and independent argument against the adaptationist point of view in the conquest of land by vertebrates.

They cannot be considered as preadaptations, either, because without the emergence of feet, they would never be used on land by vertebrates. Considering that they are independent of feet, their emergence in the evolutionary process cannot be logically viewed as preadaptations for terrestriality. To be preadaptations, they should be adapted for one use but at the same time they should also be adapted to land. If this second use is not possible at the time of emergence of these features, for lack of another feature (the feet), indispensable for the second use (but yet inexistent), no preadaptive evolutionary process is possible to be characterized. This is another new and independent argument against adaptationism in the conquest of land by vertebrates. We are going to return to the theme of preadaptation in Chapter 9 of this book.

Often neglected is the need for a functional neck for feeding (life style 11, in *fig.* 5-1). Foraging may be the most striking advantage for a land vertebrate in comparison to aquatic forms. The closer relatives of terrestrial vertebrates were large predatory aquatic animals that had little to fear in their environment. Reproduction apparently continued in water. Simply staying outside of the water, walking around, has no benefit without feeding, as reproduction could be achieved on land by other means (and in fact continued in water for a long time, apparently). Indeed, extant "amphibians", called Lissamphibia (*fig.* 1-4), such as toads and frogs (Anura), as well as tritons and salamanders (Caudata or Urodela) (*figs.* 1-1D and 1-4), usually breed in water and the tadpoles of anurans (toads and frogs) are popularly very well known. The functional neck was discussed in detail above (characters 12, 13, and 16, in *fig.* 5-1) – only character 16 is considered a possible adaptation for terrestriality, according to Blackburn (2002), but still needs to pass the criteria of Andrews *et al.* (2002), for the distinction between adaptations and exaptations. We will do this elsewhere.

Finally, a shallow-water habitat (lifestyle I in *fig.* 5-1), that apparently was the ecological niche of "†Panderichthyidae" and the first tetrapods, is believed by other researchers to benefit the emergence of terrestriality. In principle, it is easier to colonize ground from shallow water than from deep water. Thus, adaptation to shallow or marginal waters is better interpreted as an exaptation for terrestriality, because of the physical proximity of the shallow water habitat to the dry land habitat. The depressed head and body profile (character 6 in *fig.* 5-1) may not be considered related as adaptations for terrestrial life, for at least two reasons. First, because if they are already to be considered as adaptation for shallow waters, they cannot be considered to be an adaptation for terrestriality. Second, because the first tetrapod genus considered as terrestrial, †*Pederpes,* is exactly interpreted as being terrestrial because of its deep or tall, instead of low, head profile (among other skeleton features, such as the cross section of some of its leg bones, as in Clack 2002, 2012).

We have seen that 14 of the 15 studied features needed for terrestriality emerged along the evolution of osteichthyes much earlier than terrestriality. It becomes quite clear that they are exaptations for the conquest of land by vertebrates, considering the method of Blackburn (2002). Characters 1 and 2 (the emergence of lungs and the alternate movements of paired locomotor appendages) evolved before the Devonian period. This indicates that the role of the Devonian in geological and biological evolutionary events attributed to the conquest of land by vertebrates should be reviewed. Characters 2 and 3 (paired locomotor appendages with intrinsic muscles, and presence of a single skeletal element proximally to the body in these very appendages) emerged by the beginning of the Devonian (about 416 Myr). This reinforces the idea of the necessity of reviewing the role of the Devonian in influencing vertebrate terrestriality.

4 The Issue of Topology

In this book, we do not conduct a new cladistic analysis. Instead, we simply combine the results of previous researchers into a single cladogram, as shown in *fig.* 5-1. This cladogram may be called a "tree" or "topology". We consider our approach adequate for our aims, as explained at the beginning of this Chapter (paragraphs 3–10).

Some colleagues will criticize this approach, given that different researchers produce cladograms based on distinct characters and taxa. For example, they will inquire as to which characters we used in the analysis, will question why we did not include all characters here and, mainly, if the final results would not be different if these other approaches had been followed. They would furthermore wonder if a combination of all character matrices from all previous cladograms would not produce a different topology, invalidating our results with the use of the method of Blackburn (2002).

These and other reactions led us to imagine which topology would in fact invalidate our results presented in this book. On the basis of the method of Blackburn (2002), for the characters studied herein to be evaluated as representing possible adaptations for the terrestrial environment, these characters would have to have originated exactly at the same time as terrestriality. The only possible topology for this convergence of appearances of characters with terrestriality in the cladogram is that in which the terrestrial tetrapods were the sister group of the remaining sarcopterygians (or in which terrestrial tetrapods were placed more basally), as in the figures mentioned below. In other words, the cladogram of *fig.* 5-1 would have to be completely inverted. In this case characters 1–16 would all be basal in the cladogram of the Sarcopterygii, in which the †Baphetidae and the crown group would be positioned at the base, and all the remaining taxa would appear gradually, with a progressive inversion in the transformations of characters 16 to 5, which would then be interpreted as adaptations for the aquatic environment, by the method of Blackburn (2005). In *fig.* 5-22 we illustrate exactly this situation. Also, according to this method, characters 1 and 2 would continue to be interpreted as exaptations for terrestriality in vertebrates (*fig.* 5-22). Clearly, in order to make most of our interpretations of exaptations for land erroneous, Sarcopterygii would have to have originated on land. Furthermore, tetrapod sarcopterygians such as †Whatcheeriidae, †Crassigyrinidae, †Colosteidae, †*Tulerpeton*, †*Ichthyostega*, †*Acanthostega*, etc. would have to be considered progressively more adapted to the aquatic environment, starting with a terrestrial origin of Sarcopterygii (*fig.* 5-22). In *fig.* 5-22, terrestriality is indicated in bold type. We note in this cladogram that terrestriality appears as a function of apomorphies 3 to 16, which become adaptations to the terrestrial environment. Terrestriality must be considered a function of life styles I and II in this cladogram. However, terrestrialty continues to be the effect of apomorphies 1 and 2, which thus continue to be exaptations for terrestriality in vertebrates. Therefore, in an inverted cladogram, several characters must have to be interpreted as being reversions from their original states, as indicated in *figs.* 5-22 and 5-25. It is important to note that these reversions increase the number of evolutionary steps of the cladogram considerably, to the extent that it would be impossible to adopt such an inverted cladogram in publishable analyses in the specialized literature, because it would be significantly less parsimonious. Consequently, the cladogram of *fig.* 5-22 and the immediately following topologies (*figs.* 5-23

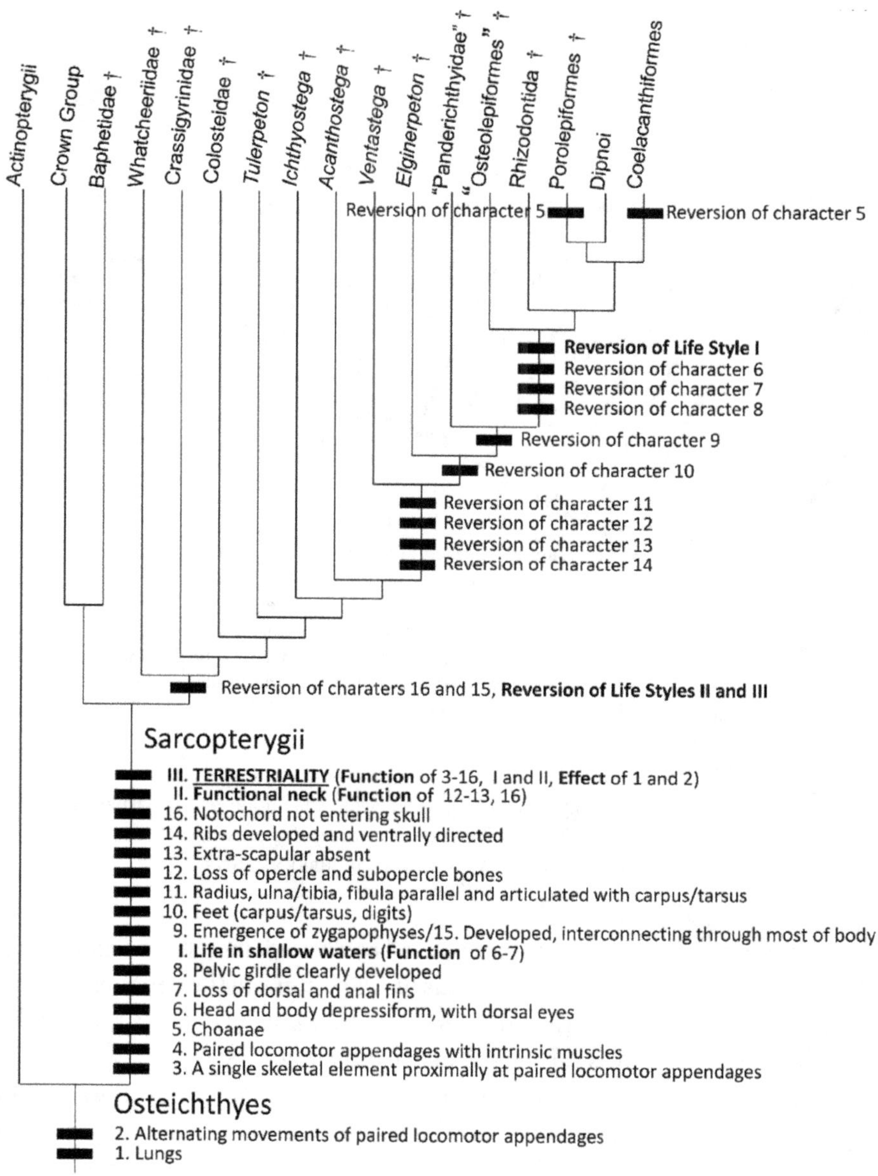

FIGURE 5-22 Cladogram of *fig.* 5-1, inverted, that is, rooted with †Baphetidae + the crown group. Even if a cladogram where the terrestrial vertebrates (included in the crown group) were more basal in position or were represented by the sister group of the remaining sarcopterygians, this would not corroborate the adaptationist point of view of the origin of terrestriality, considering the method of Blackburn (2002). This is because terrestriality appears as still being an effect of features 1 and 2. Note: this cladogram necessarily has many reversions in the studied apomorphies, contrary to the most parsimonious hypothesis. For this reason, this cladogram is cladistically untenable. Some apomorphies have been simplified in this cladogram (see *fig.* 5-1 and the text for the full description of characters). See text for further explanations.
MODIFIED FROM TRIQUES AND CHRISTOFFERSEN (2017)

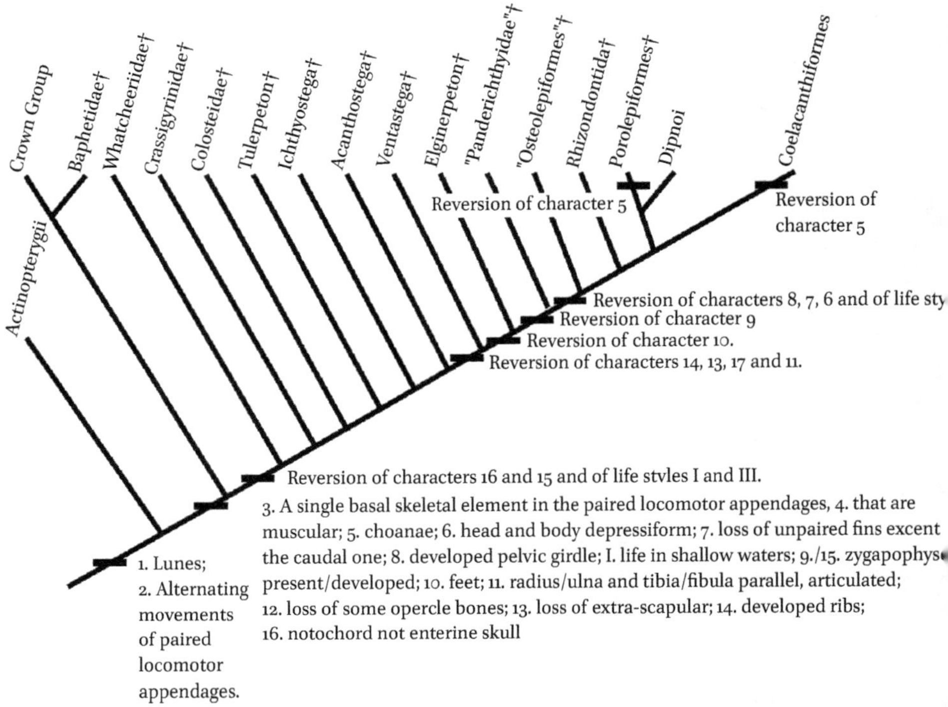

FIGURE 5-23 Simplified phylogeny based on *fig.* 5-22, given to facilitate general visualization. More details in legend of *fig.* 5-22. See text for further information.
MODIFIED FROM TRIQUES AND CHRISTOFFERSEN (2017)

to 5-25) are only working hypotheses, used solely to demonstrate what would be necessary to invalidate our proposals of exaptations in the conquest of land by vertebrates. Therefore, the criticism that our *fig.* 5-1 may not represent the best or final phylogeny is senseless, and we discard it.

A simplified and more easily visualized version of *fig.* 5-22 is presented in *fig.* 5-23, with Actinopterygii indicated a little below the studied subgroups of Sarcopterygii in order to highlight sarcoptererygians. In this way, the groups under focus become more visible in the figure. Except for the simplification, the figures are equivalent.

The propositions of new phylogenetic hypotheses may generate topologies (genealogical relationships among terminals) very different from the topology of *fig.* 5-1 and, consequently, of figures 5-22 and 5-23. But still, as only the more basal position of the crown group would invalidate the proposals of exaptations (except for apomorphies 1 and 2, as previously explained), we produced *fig.* 5-24 to indicate this situation. In this figure, the relationships among the remaining taxa of Sarcopterygii are substituted by a triangle that includes the mentioned terminals, without detailing phylogenetic relationships among

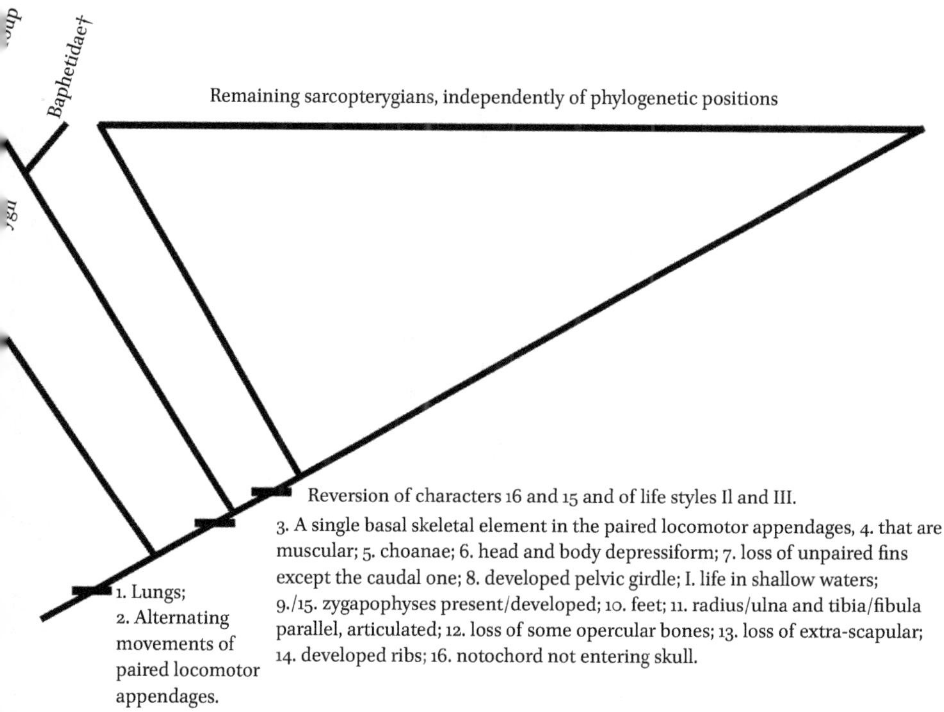

RE 5-24 Phylogeny derived from *fig.* 5-23, which differs for permitting the inclusion of any possibilities of phylogenetic relationships among any subgroups of sarcopterygians except †Baphetidae and the terrestrial vertebrates (crown group): None of the possibilities alters the fact that even the most basal position of terrestrial vertebrates would not justify the adaptationist point of view, according to the method of Blackburn (2002). This is because features 1 and 2 continue to be considered as exaptations for terrestriality in vertebrates according to this method. See text for further explanations.
MODIFIED FROM TRIQUES AND CHRISTOFFERSEN (2017)

them (these may vary freely in the future literature, without affecting our interpretations in this book).

Finally, to account for the lack of definition in the position of †Baphetidae and, consequently, of the crown group, we propose *fig.* 5-25, in order to record this possibility. Even then, this possibility requires that the crown group is positioned basally to Sarcopterygii, to permit the origin of adaptations, instead of exaptations, in the appearance of terrestriality in Tetrapoda and, consequently, in Vertebrata, by the method of Blackburn (2002). Characters 1 and 2 continue to be exaptations for terrestriality in vertebrates.

Some researchers believe that even small changes in topology may significantly alter interpretations based on cladograms. But, at least in our case, the only topology capable of significantly altering our hypotheses of exaptations is an inverted topology that is completely opposite, for several reasons, to any

FIGURE 5-25 Inverted cladogram alternative to *figs*. 5-22, 5-23 and 5-24. It illustrates the possibility of the terrestrial vertebrates (crown group) not being the sister group of †Baphetidae. A basal trichotomy leaves the position of the †Baphetidae unresolved, even though it is still relatively basal. See legend of *fig*. 5-22 for further details and text for further explanations.
MODIFIED FROM TRIQUES AND CHRISTOFFERSEN (2017)

study of phylogenetic relationships using paleontological data extensively (*figs*. 5-22, 5-23, 5-24 and 5-25). The quantity of apomorphies that insert terrestrial vertebrates away from the base of Tetrapoda in the cladogram is large. We present in the cladogram of *fig*. 5-1 only characters studied herein: characters 6 to 16, a sufficient number to sustain this affirmation. As already mentioned in this Chapter, Clack (2012) says that it is unlikely that the position of Tetrapoda within a group formed by Tetrapoda, "†Panderichthyidae" and "†Osteolepiformes", also a paraphyletic group, will change. As mentioned previously, †*Tiktaalik* is the sister group of Tetrapoda on the basis of the known fossil record, while the other two genera of "†Panderichthyidae" represent the previous genera progressively closer to Tetrapoda. Then, this large group, here denominated Tetrapodomorpha, include the terrestrial vertebrates in a derived extreme (as in *fig*. 5-1) and there is no perspective of change, from a paleontological point of view. Returning to *fig*. 5-1, we used synapomorphies 5-8, plus "life style" 1, to sustain a derived position of Tetrapoda + "†Panderichthyidae" within Tetrapodomorpha, but there are many others in the pertinent literature. Thus, other changes within this group should not change the derived position of terrestrial vertebrates within this group, which gives confidence in our analysis. More recently, Benton (2015) maintains the same topology that we use for Tetrapodomorpha, adding the term Neotetrapoda for the group we have here called "crown group ". We continue to use "crown group", in order to make it clear to the reader that this group contains the living groups of tetrapods, including the human species.

Another reason for our confidence in our hypotheses of exaptations is that, even if Coelacanthiformes and/or Dipnoi+†Porolepiformes become shown to belong to Tetrapodomorpha (or else, if some subgroup of Tetrapodomorpha comes to be excluded from this group), our hypotheses of exaptations do not change, because the terrestrial vertebrates continue to have a derived position within the larger group.

Even in the recent phylogenies of Clack *et al.* (2016) and Pardo *et al.* (2017), in which one or two terrestrial groups are taken out and placed outside of the crown group, the same characters (1–5, 7–15) are still exaptations for the crown group of terrestrial vertebrates. Some new arguments were developed by us and presented against adaptationism, in the sections on recent phylogenies, and complementary considerations, below. One more recent phylogeny (Dickson *et al.* 2020 [2021]) reaffirms the idea that Neotetrapoda (also called the crown-group Tetrapoda) is derived in Tetrapodomorpha, and again shows it as the sister group of †*Baphetes* (†Baphetidae), like our *fig*. 5-1.

Having a critical posture regarding the number of used apomorphies is of great significance. Our aim is not to conduct an exhaustive analysis of all characters with Blackburn's (2002) method. We use those characters that are most significant in morpho-functional terms, this being sufficient for the aims of this book. It is most important to note that the simple quantity of apomorphies that are analyzed does not mean that terrestriality in vertebrates may be considered to be due to microevolution. On the contrary, the appearance of terrestriality is necessarily a macroevolutionary process, because several fundamental characters for life on land are exaptations for this life style. Even if only apomorphies 1 to 2 were exaptations and all others were to represent adaptations for life on land, the process would have been macroevoltionary, because these characters are fundamental for walking on land (as was already discussed in this Chapter). Without these features, the remaining characters would not have functioned on land and would not have adapted for this life style. Thus, even if dozens or hundreds of characters turn out to be adaptations (by the method of Blackburn 2002), the process as a whole was necessarily macroevolutionary. To quantify exaptive *versus* adaptive characters would thus produce a fallacious argumentation, which we reject here. The perception that the whole process is macroevolutionary or microevolutionary is qualitative and not quantitative. An accurate perception of the characters themselves, and in their totality, is necessary. For this reason, we do not quantify our data.

It also becomes necessary to stress that the quantity of used terminals, together with a larger or smaller number of apomorphies, does not alter the macroevolutionary view on the origin of terrestriality in vertebrates, because a larger number of terminals, hierarchically organized by a sequence of apomorphies progressively less widely distributed, does not alter the derived position of terrestrial vertebrates. On the contrary, they will show a larger number of hierarchical levels existing among the origin of any of the apomorphies used on land, and terrestriality itself, corroborating our argumentation with the method of Blackburn (2002).

5 The Origins of Limbs with Digits, Walking, and Terrestriality

Clack (2012: 138) considers that the origin of tetrapods and its connection with the origin of terrestriality rests upon an argument with three parts: origin of limbs with digits, origin of walking, and origin of terrestriality. However, we consider that the focus on feet (the structure that supports the digits) is tied to the idea of microevolution and is not sufficient to explain the whole issue. The problem is only perceived when paired limbs are analyzed in their totality, as being skeleto-muscular columns with an articulated distal end (foot), as previously discussed. Furthermore, other characters are necessary on land, as seen in *fig.* 5-1, as previously discussed. When the focus becomes reduced to the suggested three parts, credit is given to microevolution as the involved process, when this is not a fact, according to the analyses we implemented in this book. The "limbs with toes", in the sense of feet, appear abruptly (as far as the paleontological record currently permits us to infer), for aquatic locomotion (or another unknown aquatic function). This is clearly a macroevolutionary process, being an exaptation for terrestriality, both in the cladogram and in geological time. Clack (2002: 138) reviews a number of previously published propositions to explain why modern fish (all actinopterygian teleosts) leave the water. She demonstrates how difficult it would be to cite any one of them as the main stimulus for tetrapods having conquered land. We consider this reasoning deceptive, because these animals (teleosts) do not have paired locomotor appendages supported by a column, nor many other characters for terrestriality, in particular characters 1,3–5, 7–16, II, precisely because there was no microevolutionary process acting in the conquest of land. The process for conquering the land environment represents, on the contrary, a rare chance event of macroevolution, as we demonstrate along this book.

Exaptations have an unrecognized importance for evolutionary biologists and derive from another concept, already perceived a long time ago. This is the perception that all phenotypic traits of an organism are integrated. Organisms are thus not disconnected collections of objects (Gould and Lewontin 1979). The implication is that if microevolution were responsible for tetrapods having conquered land, at least most characters should have evolved at about the same time. The cladogram (*fig.* 5-1) shows exactly the contrary: that these characters appeared at very different levels of the cladogram and, as will be more detailed in the following Chapter, at very different moments through geological time. Only the concept of exaptation provides a means of understanding why this is so, and provides a means of comprehending the process as a whole. Obviously, the refinement of these characters (exaptations) must have occurred along the last few millions of years after the origin of terrestriality. But this final stage was the shortest (in terms of geological time, and in terms of the evolutionary process as a whole, along a representative wide cladogram) in the evolution for the conquest of land.

For a certain character X to become progressively adapted for a certain function F1, it is necessary that several other characters are already functioning for the same function F1, in the sense of "pre-F1". Otherwise, the new function (F1) will not be able to occur, and the selective pressure for the new function will be innocuous. Microevolution is not part of the process as a whole for F1. However, if character X appears for another function (let's say, FX) and is exapted for F1 later, then the exaptation for F1 may occur, when other characters permitting the occurrence of F1 have finally evolved independently from each other, which may take dozens of millions of years to occur. Necessarily, all this "process" does not occur by "directioning toward F1" (adaptively speaking), but by several evolutionary random events. In other words, evolution is not teleological, that is, it has no direction. We must distinguish here the difference between chance in the appearance of gene alleles, and chance in the appearance and evolutionary fixation of evolutionary characters, which much later in the phylogeny of the taxonomic group may then be coopted for a new function. It is still necessary to distinguish evolutionary chance, in which several weakly connected characters must appear independently, for totally distinct functions, and finally, come to act together in order for a character X (or other) to appear, to achieve the use F1, now an effect. Ideas of natural selection, evolutionary pressures, and microevolution are perfectly correlated with much evolutionary information obtained for most taxa by evolutionary biologists and ecologists, but seem to direct perception in favor of directed change and against chance evolution, which represents a different chance event, independent of the genetic context. There are thus several evolutionary processes acting at the same time, probably in most taxa, with a higher or lower preponderance attributable to each factor in each particular case: natural selection, genetic drift, exaptation, and, maybe, other processes, still not understood.

A way of illustrating what was said in the previous paragraph is to note that, if at least one apomorphy had not appeared before or after the foot (in the cladogram of *fig.* 5-1), the conquest of land would not have occurred. The reader may imagine how, and if, we would be able to walk with 3 humeri and/or 3 femurs: there would be 3 distinct articulations in the skeletal girdle, instead of a single rounded articulation. We would thus not be able to make circular rotations with our legs, in order to move our leg backwards and forwards, and from below upwards and then down again, to give a step forward. Walking would be impossible. The same can be said about the intrinsic musculature of the paired locomotor appendages (which we popularly refer to as legs, in our case): we would be unable to articulate the knee, the elbow, the feet, in order to walk, without muscles associated with these movements. Without lungs, we would be unable to breath atmospheric air and would be unable to live on land. These characters appeared independently from each other, much before terrestriality, and were not originally connected with life on land. But, if they had not appeared independently from each other (for other functions not related to terrestriality), life on land would not have been possible for vertebrates. Evolutionary chance in these cases is the appearance of characters for other functions, without which new life styles would not be able to occur. The independence among all these characters, and their independent origins (adaptive or not), represents another great chance event. We could mention all the remaining exaptations, but will here restrict ourselves to the zygapophyses, developed to the point of interconnecting the vertebrae with each other. These interconnections emerged several steps more recently (in the cladogram) than the foot: without this additional support for the weight of the viscera – and that appeared in the aquatic environment – even locomotor limbs, completely developed, would not be able, alone, to permit walking. Thus, several chance events, totally independent of each other, permitted terrestriality in vertebrates.

6 Recent Phylogenies of Basal Tetrapoda Reveal an Additional Argument

Pardo *et al.* (2017) presented a new phylogeny of basal Tetrapoda, when studying the †aïstopod †*Lethiscus stocki*. †Aïstopoda are usually considered to be a part of Reptilomorpha, which in turn are part of the crown group Tetrapoda, currently called Neotetrapoda (*e.g.* Benton 2015: 110). †Aïstopoda are a group known from the Carboniferous and Early Permian of North America and Europe. These were snake-like animals, ranging in length from 50 mm to nearly 1 m. They had no limbs, neither limb girdles (*e.g.* Benton 2015: 102). †Aïstopods are assumed to have lost their limbs secondarily, rather than to have evolved directly from a limbless fish ancestor, according to Benton (2015: 102). The †aïstopods could have been terrestrial or semiterrestrial snake-like animals (Benton 2015: 102). †Aïstopoda are usually considered as belonging to the †Lepospondyli (also part of Reptilomorpha and sister group to †Diadectomorpha + Amniota), which may be characterized by at least five synapomorphies, three of which are secondary losses of features (Benton 2015: 110). Pardo *et al.* (2017) used micro-computed tomography of the only known specimen of †*Lethiscus* to provide new information on this enigmatic group. According to them, this information strongly challenges the consensus of phylogenetic analyses previously published on early tetrapods. Indeed, in their phylogeny, †Aïstopoda is positioned outside of the crown group Tetrapoda, a little more basally than †Whatcheeriidae (that includes the terrestrial †*Pederpes* and the aquatic †*Whatcheeria*).

The majority-rule consensus phylogeny of Pardo *et al.* (2017: 644, *fig.* 3) differs from the one we use in this book mainly in that †*Ichthyostega* is more basal than †*Acanthostega*. †Aïstopoda are the next sequential group in the cladogram, after †*Acanthostega*. In sequence, †Whatcheeriidae, then †Crassigyrinidae, and then †Colosteidae appear in the phylogeny. †Colosteidae are represented in their analysis by †*Greeherpeton* (in our phylogeny of *fig.* 5-1, the sequence toward land is the opposite: †Colosteidae, †Crassigyrinidae, and then †Whatcheeriidae). Finally, †Embolomeri (represented in their phylogeny by †*Proterogyrinus* and †*Archeria*) are taken out from Reptilomorpha and placed more basally. The †Baphetidae follow, and continue as the sister-group of the crown group Tetrapoda.

The possible consequences concerning our analysis of exaptation *versus* adaptation for terrestriality may be conflicting and are the following. (1) Terrestriality emerged more than one time in vertebrates; once in the crown group Tetrapoda + †Baphetidae + †Embolomere, another time in the now basal †Aïstopoda, and a third time in †*Pederpes* (a †whatcheeriid considered to be terrestrial by Clack (2002, 2012), as a consequence of its deep skull. (2) Developed zygapophyses (our character 15 in our *fig.* 5-1) emerged in †Aïstopoda + †Whatcheeriidae, †Crassigyrinidae, + †Colosteidae + †Embolomere + †Baphetidae + the crown group, and were then lost secondarily in the aquatic †Crassigyrinidae and †Colosteidae, which then supposedly returned to an aquatic way of life. In this case, the developed zygapophyses may be a possible adaptation for terrestriality (depending on their passing the tests for adaptation in Andrews *et al.* 2002). This is the only change for all of our interpretations above, in this Chapter. On the other hand, (3) it is still possible to consider that terrestriality emerged only once (in †Aïstopoda + †Whatcheeriidae, †Crassigyrinidae, + †Colosteidae + †Embolomere + †Baphetidae + the crown group), with three returns to an aquatic lifestyle. In this case, characters 15 and 16 are possible adaptations for terrestriality (that must be tested according to the procedures of Andrews *et al.* 2002). If the consequence (3) of the phylogeny of Pardo *et al.* (2017) is accepted, then (4) the return to water occurred three times, in †*Whatcheeria*, †Crassigyrinidae, and †Colosteidae. Paradoxically, this is an unexpected argument against adaptationism at the origin of terrestriality in vertebrates, because if terrestriality occurred by means of selective pressures for gaining ground, why are there several immediate returns (in the phylogeny) to an aquatic existence?

Further, we should consider another statement of Pardo *et al.* (2017: 645), that †*Spathicephalus* is a †baphetoid with an extremely broad skull, and is one of the most specialized stem tetrapods. According to Beaumont and Smithson (1998), the skull of †*Spathicephalus* suggests that these animals were incapable of rapid jaw closure, required for catching fish. Instead, it is proposed that this genus was a sluggish bottom-dwelling filter-feeder of small, soft invertebrates. That is, this was an aquatic †baphetid-related animal, meaning another return to water from basal terrestrial tetrapod groups. This also means that more returns to water may have occurred, indicating that it is likely more parsimonious to consider one single origin of terrestriality and several almost immediate returns (in the phylogeny) to water. Therefore, the idea that there were no selective pressures for conquering land or leaving water is strengthened.

If there were no selective pressures for leaving water or going to land, then only characters 15 and 16 were gained by chance. But even these two features must be reassessed in terms of adaptations or exaptations for terrestrial life (Andrews *et al.* 2002).

In phylogenetic systematics, it is sometimes possible to have two equally parsimonious hypotheses, one for an apomorphy to have emerged early in phylogeny, and then lost subsequently one or more times, and the other for later acquisitions of that apomorphy, with an equal number of steps (*e.g.* Wiley *et al.* 1991). The first case is called an accelerated transformation series (acronym ACCTRAN), and the second case is called a delayed transformation series (DELTRAN). Both interpretations are considered methodologically valid and the researcher is free to choose one of them. The ACCTRAN interpretation of the distribution of apomorphies in a cladogram tends to reinforce the monophyly of larger groups, and is thus usually chosen. Another reason for choosing this option is related to the concept of homology itself. The primary way to formulate a hypothesis of homology (common descendance of a feature) is by means of three basic criteria for the attribution of an hypothesis of homology for the studied feature: (A) similarity in anatomy and/or histology, in behavior, in lifestyle, and maybe also in other kinds of features, (B) the position of the feature in the body (when the character is anatomical) and (C) its function. This is called the primary homology (De Pinna 1991). After this initial proposal, the evaluation of the parsimony of distribution of all characters in the study presents us with the secondary homology, the one supported by the most parsimonious hypothesis (De Pinna 1991). In the ACCTRAN interpretation, both the developed prezygapophyses (our character 15) and terrestriality (lifestyle III) will have the most widespread distribution at their origins in the mentioned phylogeny of Pardo *et al.* (2017). Therefore, we should consider taking the ACCTRAN distribution of the two features and then their subsequent reversions, instead of considering independent origins, if we are going to follow the argument of De Pinna (1991) that the DELTRAN option puts aside the three basic criteria for primary homology. This argument represents our fourth consequence that we listed above for the phylogeny of Pardo *et al.* (2017) for the arguments based on our *fig.* 5-1.

Characters 1–5 and 7–14 of our *fig.* 5-1 continue to be interpreted as exaptations for terrestriality in vertebrates. The process as a whole continues to be interpreted as exaptive, macroevolutionary. These 13 characters were exapted together for terrestriality, possibly with the aid of some adaptations. Without the emergence of any of these 13 features for other functions in water, the collective effect of gaining ground would never have occurred (with or without correlated adaptations). We continue to interpret the origin of terrestriality in vertebrates as a chance event, now with one more argument, that of the return to water of tetrapod groups just after the origin of terrestriality in vertebrates, according to the phylogeny of Pardo *et al.* (2017), and maybe of others as well.

Clack *et al.* (2016) described five new genera of stem-tetrapods, each one with a single known species: †*Aytonerpeton*, †*Diploradus*, †*Koilops*, †*Ossirarus*, and †*Perittodus*. For †*Aytonerpeton* and †*Perittodus*, the authors described lateral line system channels in their skull bones. We interpret this as a feature of aquatic forms, because this system can only work inside water. For the other described forms, these features are not mentioned, but their absence is not stated clearly. †*Koilops* appears as the sister group of †colosteids, which are believed to be aquatic forms. Because of this, we consider this genus to be an aquatic animal. For leg bones, it was not stated clearly whether they are flat (indicating an aquatic lifestyle) or rounded in cross-section (indicating terrestrial life). Therefore, we can give support to an aquatic life for †*Aytonerpeton*, †*Koilops* and †*Perittodus*, but we cannot make any statements about the living environment of †*Diploradus* or †*Ossirarus*. A cladistic analysis is presented by Clack *et al.* (2016). A single more parsimonious tree was obtained in TNT, with an implied weight search of K = 3 (*fig.* 5a). Another strict consensus of four equally parsimonious trees was obtained from an implied weight search of K = 4 (*fig.* 5b). A third phylogeny presented is based on a Bayesian analysis, which we are not going to comment. The phylogeny of *fig.* 5a differs from ours mainly in that †colosteids are now part of the crown group (part of the †embolomeres), and †whatcheriids are more basal in position than †crassigyrinids. The terrestrial †*Pederpes* is placed outside †whatcheeriids. This first phylogeny of Clack *et al.* (2016, *fig.* 5a) implies, in terms of exaptations, that developed zygapophyses and a notochord retracted from the skull might be considered as possible adaptations to terrestriality for †*Pederpes*. Developed zygapophyses emerged independently in †*Pederpes*, on one hand, and in †whatcheeriids, on the other, but also in a third taxon, the group formed by †baphetids and the crown group Tetrapoda. Therefore, the developed zygapophyses are neither correlated to adaptation nor to exaptation for terrestriality in †whatcheeriids, because in this phylogeny this family is not in a group directly regarded to be terrestrial. Nothing else changes from our original proposal based on the method of Blackburn (2002) for the proposals of exaptations for terrestriality. However, as the aquatic †colosteids are now part of †Embolomeri, which is a part of Temnospondyli (that include living amphibians), that in turn are a part of the crown group Tetrapoda, this is an argument against adaptationism in the origin of terrestriality in vertebrates, because it represents a basal return to water in a subgroup of a group that had just conquered land. The phylogeny in *fig.* 5b differs from ours (*fig.* 5-1) in that †whatcheeriids are placed more basally than †colosteids and †crassigyrinids. †*Pederpes* is considered as part of †whatcheeriids in this phylogeny. The developed zygapophyses should be considered as an exaptation for the terrestrial habits of †*Pederpes* (because this feature is present in the whole family in which this genus is placed). The withdrawal of the notochord from the skull may be considered as an adaptation for terrestriality in this genus, pending the verification of the criteria in Andrews *et al.* (2002).

When we discussed the issue of topology above, we stated that no matter which are the arrangements between intermediary taxa, the conclusion of exaptations for terrestriality in Tetrapoda would be maintained. This is exactly what we found studying the recent phylogenies of Pardo *et al.* (2017) and Clack *et al.* (2016), except for one character. But a new argument emerged, that of the "immediate" return(s) to water (in terms of the phylogeny) of one or more subgroup(s) of otherwise terrestrial animals. If there were selective pressures for leaving water/conquer land, why would these almost immediate returns to water occur?

Triques and Christoffersen (2015: 186) already discussed the origin of terrestriality in one (or more) tetrapod groups lying relatively far from the crown group Tetrapoda. They were referring to the more distant position of the †Baphetidae in relation to the crown group (*e.g.* Laurin 2010: 162, *fig.* 7-1). According to the cladogram in Triques and Christoffersen (2015), which was slightly modified to be published in this book (*fig.* 5-1), only the withdrawal of the notochord from the skull occurred together with a functional neck. Selective forces can only be suggested at this moment, from the perspective of the method of Blackburn (2002), as stated by Triques and Christoffersen (2015: 186). In the case in which a more basal position is preferred for the †baphetids, as above mentioned, the reasoning continues to be the same. Indeed, in this case, a selectionist explanation is reinforced for character 16 (*fig.* 5-1), as the same feature would be emerging independently twice. In the three cladograms of Clack *et al.* (2016: 6, *fig.* 5), the position of †baphetids is not different from our *fig.* 5-1. However, one must consider now the phylogenetic position of †*Pederpes*, considered a terrestrial †whatcheeriid (or non-†whatcheeriid in Clack *et al.* 2016, *fig.* 5a). In the case of this genus, we would have to consider as possible adaptations not only the notochord withdrawal from the skull (character 16 in *fig.* 5-1), but also the emergence of developed zygapophyses (character 15), in the case of the mentioned cladogram, because in it, †*Pederpes* is the sister-group of †*Perittodus* + †*Ymeria*, a group that in turn is the sister group to the remaining tetrapods in this cladogram. Indeed, Triques and Chistoffersen (2015: 186) had already predicted the emergence of terrestrial taxa independently from the crown group, possibly by means of natural selection of some feature(s). We are going to discuss this again later in the paragraph. This group of three genera represents the sixth step in Tetrapoda and exaptations 1–5 and 7–14 were already present in the evolution of Tetrapoda, in Sarcopterygii, and thus in Osteichthyes themselves. It was to be expected that one single event of exaptation toward land emerged exactly at the origin of the Tetrapoda, but this did not occur, taking into account the current knowledge of the phylogeny of the Tetrapoda. Only one or at most a few groups of Tetrapoda are known to have conquered land in Tetrapoda, and one of them is the crown group. Had natural selection been acting in these animals (that is, Tetrapoda, all of them already presenting the features that exapted together toward land), at least several different conquests of land would be expected to have occurred. Thus, chance events seem more likely to have been the process that brought the animals to land, again. Therefore, the statement of Sanchez *et al.* (2016), that the transition from fish to tetrapods (maybe they were meaning the water to land transition) was arguably the most radical series of adaptive shifts in vertebrate evolutionary history, cannot be confirmed by our analyses in this book. The reason is that characters 1–5 and 7–14 in our analysis in *fig.* 5-1 were already present a long time before the first animal interpreted as being terrestrial appeared (†*Pederpes*), about 12 Myr after the beginning of the Carboniferous, and more than 25 Myr after the first tetrapod genus is known (*e.g.* Pardo *et al.* 2017, *fig.* 3). Put in cladistic terms, four steps occur between the emergence of †*Acanthostega* and of †*Pederpes* (Clack *et al.* 2016: 6, *fig.* 5a), or 6 steps according to Clack *et al.* (2016: 6, *fig.* 5b). This very large geologic time and this large number of steps in these cladograms – together with our indicated exaptations for terrestriality (characters 1–5 and 7–14) – are also sufficient to oppose the implicit idea of environmental selection in the statements of Sanchez *et al.* (2016: 408). Clack *et al.* (2016) stated that tetrapods occupied a mosaic of juxtaposed microhabitats, including ponds, swamps, streams, and floodplains, with contrasting seasonal climate, and that this varied environment persisted over the 12 Myr of the Tournaisian. The Tournaisian is a period of about 12 Myr, starting at the boundary of the Carboniferous (Pardo *et al.* 2017, *fig.* 3). Just after the Tournaisian time we have the appearance of †*Pederpes* and of other apparently terrestrial vertebrates. The very duration of the Tournaisian and its supposedly selective landscape for vertebrate terrestriality is a counter-argument against microevolution and, therefore, in favor of macroevolution toward land for vertebrates. Lungs, alternating movements of paired locomotor appendages, which already had a single proximal skeletal element and intrinsic muscles, together with internal nostrils (choanae), loss of dorsal and anal fins, clearly developed pelvic girdle, feet, radius/ulna, tibia/fibula parallel, loss of extrascapulars, as well as opercle and subopercle, and even downward directed ribs (characters 1–5 and 7–14 in *fig.* 5-1), were already present by this time. This whole set of features would not need 12 Myr only to become adapted for terrestriality. If the variations of the already present features were advantageously selected for a better adjustment to the new environment, the time requested should be much shorter. In comparison, all the adaptive transitions from land to water in Cetacea are estimated to have lasted only 8 Myr (Thewissen 2014).

The relatively recent papers we considered (Clack *et al.* 2016, Pardo *et al.* 2017) are very informative. They both present more than one phylogeny for the same set of terminals and features in each study. It is clear that there is new information on already known taxa or new taxa to be discovered in Science, but those taxa already described are incompletely known genera, having usually disarticulated skeletons. It seems to us, therefore, that it is more profitable to wait until new information completes the blanks, so that phylogenies at this level of the sarcopterygian system become stable. For the moment, the recent phylogenies such as that of Dickson *et al.* (2020 [2021]) do not change our intepretations on exaptations in the evolutivionary passage of vertebrates from water to land. Thus, exaptations and adaptations through the method of Blackburn (2002) may be reassessed in the stem tetrapods. For the present, features 1–5 and 7–14 studied in this Chapter are enough to establish exaptations and macroevolution as the conductive, main evolutionary mechanism in the gaining of ground by vertebrates.

7 Exaptation: Misunderstandings of the Concept

Exaptations have been accepted throughout every level of biological evolution, from molecular systems such as proteins and RNAs, organelles, thermoregulatory effector mechanisms up to feathers, being recognized as universal (*e.g.* Nelson *et al.* 1984, Frenkel-Pinter *et al.* 2022). However, some criticisms have been directed to the proposal of exaptation that must be addressed in this book, in the interest of clearness. We named this section as misunderstandings on exaptation, rather than criticisms, as may be encountered in the literature. One of them relates to the possible adaptive process involved in the origin of an exaptation (*e.g.* Reeve and Sherman 1993). The idea is that initially there was selection in the origin of a character for a function and, later, after this character is fully formed, a new selective process transforms the character for another function (in reality, an effect, because we are dealing with a new use for a structure that is coopted for this new use, and we are thus dealing with an exaptation). The occurrence of selection for the new use would indicate a weakness in the concept of exaptation. Yet, this is a deceptive perception, because the core of exaptation lies in the fact that the structure originated for one use, and only after being completely shaped for this use, it occurs the opportunity for cooptation (exaptation) of a structure that was ready for another use (an effect). Only after the exaptation of the feature has occurred, it (the feature) is tinkered for the new use. The second use could only be obtained if the character had appeared before, for the original and different function. The feature was already complete, before it could then be exapted. There is no adaptative continuum from the features origin for one use, to the other, as the feature was adaptively designed for one use, not for the other. For this other use (an effect), the feature already existed. Here lies the importance in the concept of exaptation. Without the first "adaptive series" by natural selection, another character would not have appeared that would be able to undergo another "adaptive series" for another function (in reality, an effect, because we are dealing with an exaptation). The chance event was the luck of there being a selective process that originated a structure (character, trait) that **could** then have been used in another way (with or without a secondary selective process, which would never have occurred, if the first selective process had not already occurred).

The most important perception is that the same selective pressure can select more than one different feature that fits with it. One of these features fortuitously can have a secondary use (and eventually turn into an exaptation, at a less inclusive hierarchical level). The others, not. For instance, selection for living on the bottom selected many different vertebrate groups. Consider the †placoderm genus †*Phyllolepis*, the ostariophysan (teleost) banjo catfishes, and the chondrichthyan rays. The rays have much enlarged pectoral fins that undulate for propulsion of the body. Only manta rays live just under the sea surface (as a filter feeder), and undulation of enlarged pectoral fins, together with the flattened body, are exaptations for pelagic life right below surface. It is clear that natural selection for bottom living shaped several different features for living on the bottom and that one of them (together with the depressiform body) could be exapted for surface-dwelling (the pectoral fin undulation for propulsion). Therefore, this is a case to show that the same selective pressures may generate very different features, one of which is exaptable for another lifestyle.

A simple example are the feet. They were selected as an advance for aquatic use and could be exapted for land use. Other aquatic vertebrates "responded" to the selective pressures of the aquatic environment generating different features. For instance, in derived actinopterygians there seems to have occurred a reduction in the number of pectoral fin radials. Teleosts usually have only three radials (*e.g.* Weitzman 1962, among a large number of other papers). Two rows can be seen in basal forms like cladistian actinopterygians, such as †*Polypterus* (*figs.* 1-3 and 5-13). The paired appendage skeletons in the sarcopterygian †*Eusthenopteron* (*figs.* 5-2 and 5-12) are shortened from the same original elongate skeleton that gave origin to feet (the plesiomorphic condition seen in *fig.* 5-11). Therefore, the same or similar selective pressures led to different features, one of which by chance could be used on land, as an exaptation, later in geologic time.

The example described above also shows that Reeve and Sherman (1993) were mistaken when they stated that there is a new adaptive series for the second use on land. The skeletal arrangement of feet known from aquatic to terrestrial forms is regarded only to decrease the number of fingers (*e.g.* Clack 2002, 2012, Pough *et al.* 2005, among many others), not to change the original feet totally, as apparently happened in the case of the avian and pterosaurian wings. A slight degree of selection for adjustments to the new use was already described by the proponents of exaptation (Gould and Vrba 1982). This so-called secondary adaptation is by no means a way to originate a new feature, however, as one can see in Gould and Vrba (1982).

The first adaptive series may have generated any structure, efficient for the adaptive necessity of the species, not cooptable for a particular effect, as in the above example of the origin of feet. It is this possibility that has escaped the perception of the critics of exaptation.

Another criticism of these authors (Reeve and Sherman 1993) is that it is impossible to discern exactly when an adaptation ends, and an exaptation begins, as Gould and Vrba (1982) do not specify (according to Reeve and Sherman 1993) when a current function of a trait (character) differs from its original function to be classified as an exaptation. But in this book, we have had no difficulty in showing aquatic and terrestrial uses. Even if there are situations in which this difficulty exists, there is no reason to ban the use of the concept of exaptation from evolutionary biology. The method of Blackburn (2002) is very useful to discriminate between adaptation and exaptation, in many cases at least.

Still another criticism of Reeve and Sherman (1993) is that practically all characters are exaptations, when one studies them backwards from a phylogeny of the group. In our view, this still remains to be demonstrated. For example, in a key to the species of "tetras" (small fish appreciated by aquarists) in Tetragonopterinae (*e.g.* Géry 1977), the number of characters of the external anatomy which are used is very large, as are the number of "states" of these characters (in this case, differences in the characters of several species). These character states appear along the evolution of the group, without there being a clear change in the function of the character in different species, genera, and groups of genera. In this way, these characters are very unlikely to represent exaptations. They represent adaptations or the result of genetic drift. This has still not been the object of specific studies, given the absence of broad phylogenies of this subfamily of Characidae. There is also a lack of professional interest in discriminating between adaptations and exaptations in Characiformes (a group of freshwater teleosts from Africa and the Neotropical Region, that includes the tetras, neons, piranhas, trahiras, South American darters, pencilfishes, Amazonian hatchetfishes, African freshwater tiger fishes, etc). Clearly, the method of Blackburn (2002) is of great help in this situation. It permits us to distinguish which are adaptations and which are exaptations. Other criticisms of Reeve and Sherman (1993), on the difficulties of distinguishing adaptations from exaptations, such as the possible fastness in the transition of the "selective regimes ", which make such distinctions more difficult, should be resolved by well supported phylogenies, and by the use of the method of Blackburn (2002).

Some characters are clearly unlikely to be exaptations. For example, which character, trait, or structure originated lungs? Another example of structures unlikely to be of exaptive origin are the horns of several groups of mammals, functioning in defense (rhinoceros) or in sexual displays (deers, elks), for example.

In any case, even if all characters represent exaptations at some evolutionary level, an important alternative interpretative possibility is presented, against the notion that the environment influences the evolution of all and any groups. The importance of natural selection would be largely reduced. At the same time, chance – as here defined – becomes established for each and every character. This becomes another important consequence established here.

Another frequent mistake is the belief that a character appears and then it is gradually modified in its subgroups, occupying vague ecological niches. In reality, when exaptation occurs, the initial origin of a character occurs by a certain adaptive process (or by a neutral process). Without the presence of that structure, other functions would not be able to appear (either by natural selection or by other non-selective processes), simply because the structure would never have existed in order to be changed later. Again, we note the action of chance, as the second event of adaptive changes was only able to occur because the structure (character, trait) appeared by means of a selective process. First, a character or trait originates, then the character is used in another way, occasionally originating another selective process (different from the first selective process), which would not have occurred in case the first selective process had not initially selected the structure (once other, different features could have been the result of the original selective pressure). The evolutionary process is not continuous, because the second selective process occurs not to change the structure for a second function (because the structure is already available and useful for the second use), but this second selective process (for refinement of the already existing feature to the second use) would not exist without the origin of the structure by means of the first selective process and, by a simple accident, be also useful for a second, different use. Thus, it is a mistake to interpret that the feature erected by one selective pressure is recreated completely anew by a second selective pressure for a second use. The feature is ready, after the primary adaptive selection. It is only slightly adjusted by a secondary selection to better fit the second use, as an exaptation.

The foot in our interpretation possibly originated in order to enhance the performance of the fin as a hydrofoil to help direct movements within the water, and maybe to help functioning as a paddle, to maintain position when the animal is steady in the environment (as fishes do in an aquarium). It may also be a structure that appeared, hypothetically by natural selection, to improve these aquatic functions. These functions probably were already present in the previous state of these paired locomotor appendages (state "without a foot"). This state is structurally totally different in the paired pectoral and pelvic fins of Actinopterygii (*fig.* 5-15), exclusively rayed, and with rounded, instead of elongate bases. Therefore, actinopterygian fishes attain the same function without a "foot" or any structure similar to a foot, by means of delicate rays (and sometimes spines derived from rays). The chance event that led us to walk on land is the origin of a foot. However, such a structure could have not appeared, as it is unnecessary for actinopterygians, chondrichthyans, †placoderms, †acanthodians and the fusiform sarcopterigian †*Eusthenopteron*. Elongated rays could have appeared in the place of the foot, but then, by evolutionary chance alone, this did not happen. Note that the "osteolepiform" †*Euthenopteron* presents about the same situation occurring in Actinopterygii, but belongs to the Sarcopterygii (*figs*. 5-2 and 5-12). In other words, despite being a tetrapodomorph rhipidistian (*fig*. 5-1), despite being close to the Tetrapoda that present a foot, †*Euthenopteron* does not have a foot. The failure to understand that natural selection does not select necessarily exactly the same structure in different taxa, nor only a single structure for a particular function, prevents the understanding of the role of exaptation and macroevolution. Natural selection, then, effectively may select distinct characters in distinct taxa, in response to the same selective pressures. As DNA varies in an aleatory way, one should not expect the same variation would occur in different taxa. Therefore, natural selection probably developed, or may have developed, different but equally efficient features for the same selective pressure. One or more of these characters may be coopted for another function, representing an exaptation, one of the bases of macroevolution. In case only the most efficient structure had been selected, possibly the foot would not have been selected, because the paired fins of †*Eusthenopteron* and of the actinopterygians are apparently more efficient (because these taxa are more abundant in aquatic environments, both in number of species and in biomass, both in fresh water as in salt water). In this case, a foot would not have appeared, and we would not have been here. This is exaptation and macroevolution. This is the kind of chance event we would like to make clear to the reader.

Researchers finding it difficult to understand the concept of exaptation, view continuity in the evolutionary process. However, in reality, the utilization of a structure has changed from one use to another, but the structure itself is the same (occasionally only slightly tinkered for the new use). The second selective process is different, even because the first was necessarily a selective process for the origin of the structure. Later, this structure would have undergone a shift in use, and only after the shift occurs, may a different selective process originate, that does not transform the structure into a different one, but only tinkers it to perform better for the new use (or effect, as Gould and Vrba 1982 prefer). This other selective process for another function would never have occurred if the structure had not been formed in the first place, for another use. The passage from an original function to a totally new function represents the action of chance in evolution.

In our view, the mistake of viewing the evolutionary process as continuous in the conquest of land becomes clear by the fact that terrestrial vertebrates form a monophyletic group. A similar selective process, which acts on non-related taxonomical groups (such as Chondrichthyes, Actinopterygii, and Sarcopterygii), or even on closely related groups (such as in subgroups of Sarcopterygii), led to very different consequences. Only in Sarcopterygii did features emerge that, later, could jointly be exapted for terrestrial life. Had it been strictly a continuous process of natural selection toward land, several non-sarcopterygians would be expected to have conquered land.

Totally distinct structures may be selected to better adapt an organism to the same selective pressures and, consequently, environmental conditions do not lead to a predetermined result. This "final result" will only be perceived as such, because taxa appear to fit into a certain life style. The interpretation of several researchers is that these taxa originated by selective pressures, and thus that they evolved into several perceived shapes. In reality, a cladogram may indicate the occurrence of several exaptations instead of adaptations and, at the same time, may indicate the monophyly of a group that would be expected to be polyphyletic. The structures for walking and for life on land are thus not a direct or necessary consequence of evolutionary scenarios in the Devonian, or of any other temporal time division. Life on earth results from chance events in the cooptation of several preexisting characters for terrestriality.

Thus, the conquest of land did not occur only because there was change in the terminal structure of paired locomotor appendages (the fins) of aquatic animals, into the so-called foot, as some researchers envision. The foot continued with an aquatic function for millions of years. For the foot to be functional on land, several other characters (apomorphies that appear along the cladogram) had to have appeared independently, both before its origin, and also after the appearance of the foot (*fig.* 5-1). Until the set of characters was completed, the foot could not have been useful on land. This is chance in evolution; this is macroevolution. Clearly, the whole set of characters could never have been sufficiently complete for terrestriality to have appeared in the vertebrate animals.

Characters necessary for terrestriality, having appeared for other uses, were exapted for life on land, a process we call macroevolution. Certainly, at this point, natural selection may have acted to trim some of these exaptations for their new uses, in a process termed secondary adaptation by Gould and Vrba (1982). This second adaptation series could only have occurred when the foot was already in place, for an aquatic function, and the second adaptive series could have operated on other exaptations, as we indicated above, only after the beginning of their use on land. The beauty in the process lies in that chance was responsible for the origin of a foot for one function, permitting another totally different function (in reality, an effect) to occur. Macroevolution, at least in our redefinition, refers to an interruption of one adaptive series for one function, toward another secondary adaptive series on the effect of the structures originating in the first adaptive series.

We consider that the term and concept "preadaptation" is, or may be, part of the misunderstandings on the concept of exaptation. Certainly, the first may be used in an attempt to invalidate the second, and we will work against this attempt. We are going to deal with this issue in Chapter 9.

Misinterpretations of the concept of exaptation occur occasionally through the literature using this concept. For instance, McLennan (2008: 249) stated that Gould and Vrba (1982: 11) would have exchanged preadaptations for exaptations but they indeed stated that preadaptation is an exaptation considered before the fact. Another example comes from research on human evolution, more specifically on the origin of consciousness. Tattersall (2004: 25) states that any novelty has to arise spontaneously as an exaptation, a structure existing independently of any new function for which it might later be coopted. This statement is not in conformity with the concept of exaptation. Exaptation may be either a process or the consequence of the process. Therefore, no features arise spontaneously as exaptations. A feature may emerge as a by-product of the emergence of the evolution of other feature(s). This case was termed "spandrel" by Gould and Lewontin (1979). Only later would this feature be exapted by an effect, becoming then an exaptation for that use. A feature may emerge as a selected character for some use (for example, feet for better maneuvering in water, according to us, in this book) and later be co-opted for another use or effect (feet for walking, for instance). Therefore, features do not arise spontaneously as an exaptation. To avoid such misleading uses or interpretations, it would be prudent to refer to the original publication on the term exaptation (Gould and Vrba 1982), for a precise definition. In order to understand the whole evolutionary process regarded to represent an exaptation, we recommend referring also to Gould and Lewontin (1979). The idea of exaptation from spandrels (by-products) helps us to better correlate this associated concept with the concept of exaptation. It also helps us to understand the original criticism of adaptationism. In any case, the feature must appear first, in order to be later exapted at a less inclusive hierarchical level of the phylogeny, instead of arising as an exaptation, as stated by Tattersall (2004: 25). In that case, it would be an adaptation.

Some authors are simply not aware of the existence of Blackburn´s (2002) method, and suppose that there is a "non-operationality objection" to the use of the concept of exaptation. By itself, this is not a misunderstanding of the concept, but a source of questioning or criticism that may spread in a way that could lead some researchers back to adaptationism. For instance, almost a decade after Blackburn's (2002) publication, Pievani and Serrelli (2011) still discuss this "objection". Of course, this is a matter of ill-propagation of ideas and methods on exaptation. Not only Blackburn´s (2002) method is not known by Pievani and Serrelli (2011), but neither the criticism and propositions from Andrews *et al.* (2002) seem to be known by them, and by most researchers on human evolution. These works remain uncited, at least to 2011, as is possible to see by consulting the literature on which Pievani and Serrelli (2011) developed their own arguments, namely Shimizu and Macho (2007), Tattersall (2004), and Kivell and Begun (2007), all published well after Blackburn (2002). Blackburn´s (2002) method was published in a well-known and well evaluated magazine in the area of brain and behavior Science, basically the same or at least in the correlated area of anthropological, human evolutionary Sciences. Considering the unfamiliarity of this method of Blackburn, it is possible that many other researchers rely on the so called non-operationality objection, indeed non-existent nowadays. Hopefully our job will partially fulfil this gap in general understanding.

As a consequence of Pievani and Serrelli's (2011) oversight, they present their six directions of research that can effectively contribute to operationally discern exaptation from adaptation, case by case. We are going to discuss each one and refute them, subsequently. Their first direction is to employ biomechanical studies and models to analyze function. However, this is mere consistency, as already criticized by Andrews *et al.* (2002). Of course, non-anatomical features do not fit this direction. Non-morphological exaptations cannot be discerned from adaptations with this direction. The second direction put by the authors is to test for effective correspondence between structure and function in living and fossil species. The adaptive hypotheses need to be supported by consistency between the considered structure and function. They complement their statement by adding that, if the distribution of a trait is random regarding a behavior and lifestyle, an adaptive hypothesis is undoubtedly weakened, and a "convergence approach" (quotation marks by Pievani and Serrelli 2011: 16) can potentially give strong support to an adaptive hypothesis. We consider that a clearer test is that of Blackburn (2002), because even a non-random distribution of a feature may be shown to be an exaptation, while a random distribution of a feature may be demonstrated to be an exaptation. The third test Pievani and Serrelli (2011: 16) bring forward is to explore multiple functions of a structure: functions that are now secondary (and may be underestimated by researchers) could have been primary in the evolutionary past. We do not see this proposition as a test. As we discussed regarding character 10 in our *fig.* 5-1 – the foot – the original function may have been simply to improve the function of the limbs as a hydrofoil, and this was not considered by previous authors. But our supposition itself is not a test. Neither did we use this as a test for the feature, but as an application of Blackburn's method. The exaptive nature of the feet for terrestrially was clearly shown in our *fig.* 5-1. The fourth proposition of the authors is to explore structural alternatives for the considered functions. They believe that the existence of actual structural alternatives for the same function can weaken the structure-function correspondence necessary for a strictly adaptive hypothesis. But, as we have pointed out for the vertebral zygapophyses (feature number 9 in our *fig.* 5-1), and in regard to body rigidity, the emergence of flat, widened ribs in †ichthyostegids for the same function may suggest that selective forces were active in this direction. That is, we see the opposite result for the fourth test of Pievani and Serrelli (2011: 16). In other words, each feature with the same function would be an independent adaptation for a given pressure. The fifth proposal of the exaptation/adaptation test put forth by Pievani and Serrelli (2011: 17) is to enlarge the phylogenetic context and improve knowledge, according to them. They state that an accurate account of the occurrence of the considered trait in a complete phylogenetic tree is the foundation for understanding the balance between phylogenetic information (common descent) and functional information (features relative to environment and lifestyle), carried by the trait. By the expression "features relative to environment and lifestyle", we believe the authors mean what we regard as "function" for adaptation and "effect" for exaptation. If so, they do not explain how the distribution of a feature and function/effect in the cladogram should be interpreted, as indeed was done by Blackburn (2002). On the other hand, they rely on "random distribution" (quotation marks are ours) that, if found, could be either accepted or used as a basis for further research in the same way as their second test. We discussed the second test of Pievani and Serrelli above and discarded it. With the same arguments, we discard this fifth test of theirs. To improve knowledge concerning ontogenetic and developmental processes underlying the structure is the sixth test of Pievani and Serrelli (2011: 17). They add that the adaptationist view often relies on the assumption of direct genetic control of structures. These authors state that there are studies that employ a more integrated view of ontogeny and development. According to them, Shimizu and Macho's (2007) hypotheses concerning "morphodynamic processes and mechanical forces" (quotation marks of Pievani and Serrelli) show that "genetic networks" (quotation marks are theirs) are only one level of factors acting in pre-natal development, which thus requires multiple levels of analysis. They see a recurrence of the idea and argument on the increasing amount of information as a kind of revolutionary event, which is capable of changing perception on human thinking. We disagree with this point of view. It is not a larger amount of information that changes human thinking. It is the change in human thinking, generating new premises, that further changes human thinking, and thus changes the way the new information is understood and sought. Neither the knowledge on DNA nor the existence of information from an electronic microscopy, nor even the current knowledge on zoological species numbers, were necessary for Darwin and Wallace (1858) to publish their ideas on the origin of species by means of natural selection. The criticism on adaptative thinking by Williams appeared in 1966, with the statement that the proposal of adaptation is a burden of proof for the researcher. Also, Williams (1966) created the whole concept of exaptation, only not providing a name for the concept. This name was created only much later, by Gould and Vrba (1982), not because new information was published, but because Gould and Vrba perceived the importance of the concept of Williams, and the need for a name. Finally, the method of Blackburn (2002) does not demand more information to be applied with sharp results, or an integrative approach of information. New information and integrative approaches are very welcome, but they do not constitute a test for hypotheses of adaptation or exaptation by themselves.

Coddington's (1988) cladistic tests of adaptational hypotheses are not intended to differentiate adaptation from exaptation in a cladogram. This is not a misunderstanding of the concept of exaptation, but we inserted this citation here for the sake of propagation of knowledge. We could have inserted it in any other position in this Chapter. Greene (1986 *ap.* Baum and Larson 1991: 3) invokes the criterion of historical genesis, by testing for phylogenetic congruence between the origin of a trait and the measured performance advantage. For a trait to qualify as an adaptation, its origin must coincide with that of the performance advantage. However, the performance advantage evolved in penguins much later than did their wings. Therefore, wings in penguins constitute an exaptation for swimming (Greene 1986 *ap.* Baum and Larson 1991: 3). Thus, the interpretation of Greene seems to be the same as that of Blackburn (2002). Baum and Larson (1991) created a phylogenetic method for studying character macroevolution. Indeed, they test hypotheses of adaptation and discriminate adaptation from alternative causal explanations of character evolution, according to the authors. Baum and Larson (1991: 1) state that the criterion of historical genesis is applied using the concept of "selective regime" (quotation marks is theirs), which is said to be the aggregation of organismal and environmental factors that combine to determine the expected action of natural selection on actual and potential character variation. The method seems to be the following: an adaptation that arose in a lineage having the same selective regime as the focal taxon constitutes an "adaptation", whereas one that arose in a lineage having a different selective regime constitutes an "exaptation" (quotation marks theirs, in both cases). For us, this seems to be misleading, because the method does not clearly discriminate whether the feature emerged before or after its function, as does the method of Blackburn (2002). That is, the method of Baum and Larson (1991) does not follow the criterion on historical origin. At least, the method of Blackburn (2002) makes historical origin more direct and clearer. Also, for us, it seems hard to know exactly whether the selective regime acted at the origin of the feature or at some other moment, because this information is not necessarily available from the fossil record. For the present time, the information may be available, but it is not necessarily the same as the one at the origin of the monophyletic group in consideration. Although we do not regard the method of Baum and Larson (1991) as a misinterpretation of the concept of exaptation, we are including it in this section of our book, because the method may include uncertainties, as we pointed out. We do not aim to analyze further the method of Baum and Larson (1991), but it is noteworthy that other methods were proposed, besides that of Blackburn´s.

8 Complementary Considerations

Clack *et al.* (2016) consider positioning †*Pederpes* in †Whatcheeriidae. This represents another hypothesis for the origin of terrestriality. Some considerations will be made in a subsequent Chapter. Firstly, the previous emergence of most of the studied apomorphies as exaptations for terrestriality can be observed from *fig.* 5-1. After the emergence of characters 1–5 and 7–14 (including feet), it would be expected that this set of features would be enough for a collective exaptation toward land. At the same time, it would be expected that few adaptations would be necessary, considering the presence of all these 13 features exaptable for land. Secondly, one must note that, although †*Pederpes* may be considered a terrestrial †Whatcheeriidae, this terrestrial evolutionary line did not leave any other known terrestrial evolutionary line (descendants), as far as known. Pardo *et al.* (2017: 644, *fig.* 3) present information not only on the phylogenetic relationships among the taxa they studied, but also on the distribution of these taxa in geological time. Together with †*Lethiscus*, †*Pederpes* is the first known terrestrial vertebrate. However, if †*Lethiscus* is part of the possibly terrestrial †Aïstopoda, a lineage that lasted at least until the end of the Carboniferous, †*Pederpes* or †Whatcheeriidae are not known to have lasted more than some 10 Myr (348 to 338 Myr approximately), in a time where no other terrestrial group of vertebrates is known except for the †Aïstopoda. Thus, although no terrestrial vertebrate competitors are known for †*Pederpes*, this genus and its lineage did not leave terrestrial descendants. This fact is hard to explain in an adaptationist view of evolution. Thirdly, it is necessary to take into account the possibility of †Whatcheeriidae being a polyphyletic group, based on leg bones morphology, of the aquatic type of †*Whatcheeria*, and of the terrestrial type in †*Pederpes* (Clack 2002, 2012), as well as the variable position of the genus in Clack *et al.* (2016). Therefore, a final resolution of †*Pederpes* as part of Neotetrapoda still seems possible (in the crown group Tetrapoda).

In Clack *et al.* (2016), †*Pederpes* appears as part of †Whatcheeriidae, or groups with other genera. In a cladogram of Clack *et al.* (2016 *fig.* 5c), †*Pederpes* is a little derived in †Whatcheeriidae, a family that presents interconnecting zygapophyses (feature 15 in our *fig.* 5-1). In this case, this feature continues to be considered an exaptation for terrestriality in †*Pederpes*. In the other cases, however, feature 15 may represent an adaptation for terrestriality in this genus.

CHAPTER 6

Geological Time, Morphology, and Ecology

Geological time has not been used previously as a method to distinguish exaptations from adaptations, or macroevolution from microevolution. Yet, geological time may be useful for discussing these evolutionary matters, not only intuitively, but also in logical terms. Some correlations may at least be suggested for arguing in favor of one or another possible evolutionary process. How long do natural selection, selective pressures, and microevolution take to produce or to shape a character for a particular function? One character may adapt faster or slower than another, so this factor always permits the questioning of the use of geological time as an indicator of micro- or macroevolution. However, when geological time for the refinement of one character is seen to be very much longer than for other characters for other functions, macroevolution may be suggested as the evolutionary process involved.

Consider adaptation by natural selection. For one feature variation occurring in a species to be selected positively it should be somehow advantageous from the parent's generation to that of the next generation. Therefore, it is logical to expect that a feature variation should be fixed in a species with limited geographic distribution for hundreds or thousands of generations. Considering a species in which individuals live about 50 years, it seems reasonable to suppose that in about 5 to 10 Myr a variation would be fixed in the species. Thus, a time space above this period is long enough to at least raise questions on the proposition that the feature variation is or was an adaptation for a given environment. More significantly, the features we discuss here are ready, complete as a useful feature, and therefore functional. Thus, we are indeed discussing the time needed for their tinkering for a different, secondary use as exaptations. The time expected for the secondary adaptation to become an effect should be even shorter.

To be more specific, let us consider the work of Nilsson and Pelger (1994) on a "pessimistic estimate of the time required for an eye to evolve", as the authors titled their paper on a computer model study that, according to Ridley (2004: 260), illustrates the power of Darwin's argument. Nilsson and Pelger (1994) state that theoretical considerations of eye design allow us to find routes along which the optical structures of eyes may have evolved. If selection constantly favors an increase in the amount of detectable spatial information, a light-sensitive patch will gradually turn into a focused lens eye through continuous small improvements of design (Nilsson and Pelger 1994). An upper limit for the

 | DOI:10.1163/9789004684911_007

number of generations required for the complete transformation can be calculated with a minimum of assumptions (Nilsson and Pelger 1994). Even with a consistently pessimistic approach the time required becomes amazingly short: only a few hundred thousand years (Nilsson and Pelger 1994). Ridley (2004: 261) summarizes the findings of Nilsson and Pelger (1994) as follows: From the initial simple stage, Nilsson and Pelger allowed the shape of the model eye to change at random, in steps of no more than 1% change at a time. One percent is a small change and fits in with the idea that adaptive evolution proceeds in small, gradual stages. The model eye then evolved in the computer, with each new generation formed from the optically superior eyes in the previous generation. Changes that made the optics worse were rejected, just as selection would reject them in nature. Ridley (2004: 262) ends his summary stating that the complete evolution of an eye similar to that of a vertebrate or octopus took about 2,000 steps. What had looked like an impossibility, actually turns out to be possible in a short interval of time. "With a generation time of 1 year, the evolution of an eye from a rudimentary beginning would take less than 0.5 million years. Far from being difficult to evolve, the model shows that it is rather easy" (Ridley 2004: 262). Converting this rate for that in which individuals live 50 years, as we considered above, we would have 25 Myr. However, the modeled eye studied by Nilsson and Pelger (1994) started with three layers of flat cells: upper transparent cells, light sensitive cells, and a dark pigmented bottom cell layer. The cases we examined in this book include only completely formed characters, which would only need to be tinkered for use on land. Therefore, our "calculation" that resulted in 5 to 10 Myr only, for the adaptation for use on land of an already complete character, seems reasonable. Further, we could follow Nilsson and Pelger (1994) for a generation time of 1 year, because after gonadal maturation, the adults that were supposed by us to live 50 years, would spawn every year for more than a couple of dozen years, leading to an evolutionary rate of the studied characters in *fig.* 5-1 closer to the findings of these authors. In any case, our suggested time of evolution seems to be reasonable.

Thus, how fast is evolution? This question of Harmon (2014: 567) is answered by himself, stating that Darwin's finches speciated so rapidly that it is difficult to find out which were their phyletic relationships to each other, a nearly simultaneous explosion of species having occurred. But he goes further, showing that speciation may be very slow, calling attention to the coelacanth, which has undergone little or no evolutionary change. Harmon recognizes that these seem to be exceptional cases. According to Harmon (2014: 567–568), rates of evolution may play a key role in several controversies in evolutionary biology, including punctuated equilibrium, and also on debates about adaptive radiations, for instance. Rates are calculated on the average rates of speciation and extinction per lineage per million years, and these rates vary tremendously through the phylogeny of taxa (Harmon 2014: 568). This is a mathematical model, which means that they have their own limitations and do not describe the rate of mutation or evolution itself, completely. It is more suitable for evaluation of speciation and data comparison (Harmon 2014: 568). Therefore, it cannot be described as a definitive way to determine whether a trait should be said to evolve slowly or fast. According to Harmon (2014: 570), lineages also have traits that evolve and change through time. Their rate estimation is done in two ways. Both rely on fossil data. Time may be used to calculate a rate or, alternatively, to measure the features of the trait itself, but now in recent groups. The differences found were then evaluated by way of looking at a common ancestor at some known point in time. Both methods yield average rates over time, which may underestimate the amount of change that occurs in each generation (Harmon 2014: 568). It is implicit that each taxon or each trait may have, and usually has, its own evolutionary rate, and that the calculations are made on fossil data. So, it is not possible to tell exactly how fast or how slow an evolutionary trait *should* be. Yet, these calculations are not general enough to tell adaptive versus exaptive processes apart. In any case, (Harmon 2014: 571) states that if the observed trait is under directional selection, we expect to see a much faster change than that expected under genetic drift. High rates of both speciation and trait evolution are key components of the concept of adaptive radiations, and some clades have been shown to evolve faster than others (Harmon 2014: 572). It is thus possible for us to make our own considerations on the matter.

In comparison, we may stress that the Carboniferous lasted for 60–65 million years (*fig.* 2-1). This represented sufficient time for (1) the appearance of clearly terrestrial vertebrates, for (2) an extensive diversification of batrachomorphs (non-amniotes) and several reptilomorph non-amniotes, and (3) for the appearance of Amniota and several of their lineages, including the Synapsida (which include Mammalia), as may be seen in textbooks (*e.g.* Pough *et al.* 2005: 198, *fig.* 9-1; Benton 2015: 109–110, 123–124, boxes 4.5 and 5.1; Pough and Janis 2019: 162, *fig.* 10-1). The following question thus emerges: in an equivalent time period, how do the studied structures behave adaptively? For example, how much geological time did the already completely structured feet require to change their use from an aquatic to a terrestrial environment? A short time would clearly indicate adaptation, because, when there are adaptive advantages, the structure will be selected, and this process can only be geologically "quick" (because advantageous). However, when time since the appearance of a complete feature until its adaptation for a new function is prolonged (for a new effect, as one would say for an exaptation), how can we argue that there was an adaptive "advantage"? In other words, a long time could indicate an exaptive component involved in the process. A possible criticism of this view would be that "long" and "short" evolutionary times, and "rapid" or "slow" evolutionary rates, are subjective evaluations, and subjectivity must be avoided in Science. However, the enormous amount of evolutionary diversification of terrestrial and aquatic tetrapods in the Carboniferous, in comparison with the times necessary for the feet to develop from aquatic to terrestrial functions, and the evolution of many other characters may be evaluated comparatively, maybe thus eliminating the subjective aspect, and making it possible to argue for or against microevolution or macroevolution. Although the subject remains polemical, we will present our argumentation below, because the comparative terms are, most often, clear.

Basically, the longer the time necessary for the origin of a feature by selection, the less adaptation is justified. In terms of discussing the choice between hypotheses of adaptation or exaptation, we should remember that the features are already complete and functional for use in water. Then, the longer the time for adaptation to terrestriality, the less justified is the hypothesis of adaptation. That is, the case involves a larger likelihood of being an exaptation process, instead of an adaptive process toward terrestriality.

In order to correlate the evolution of characters with the geological scale (*figs.* 2-1 and 5-1), we use minimum ages (those presently known) for taxa: 360 Myr for the conquest of the terrestrial environment by Tetrapoda, coinciding with the origin of the Carboniferous period (*e.g.* Clack 2012: 6, *fig.* 1-2); 395 Myr for Tetrapoda (Niedzwiedzki *et al.* 2010); 410 Myr for Rhipidistia (Janvier, 1996: 24, *fig.* 1-23); 416 Myr for Sarcopterygii (Zhu and Yu 2009), although Clack (2012: 60, *fig.* 3-2) indicates the minimum age of Actinopterygii as being around 420 Myr, in the Silurian period, and she indicates that the "fish with lobed fins" would have the same age; 440 Myr for Osteichthyes, considering that Janvier (1996: 24, 181–182; *fig.* 1-23) attributes this age to †Acanthodii, the sister group of the Osteichthyes (although Pough *et al.* [2005, *fig.* 3-14] indicate the age of †Acanthodii as being 450 Myr, as well as Pough and Janis (2019: 96: *fig.* 6-1); 408 Myr for Dipnoi (Janvier 1996: 24, 181–182; *fig.* 1-23). All these minimum ages are based on the fossil record, except that suggested for terrestriality in Tetrapoda. The first known terrestrial vertebrate is †*Pederpes*, which has a foot apparently adapted for terrestrial locomotion, being of "Ivorian" age in the Carboniferous period, 348–344 Myr (*e.g.* Clack 2002, 2012). Thus, the "age" of the beginning of terrestriality in Tetrapoda may have been, in fact, around 360 Myr, at the beginning of the Carboniferous, because it is not unlikely that terrestrial tetrapods older than †*Pederpes* will be found in the future, closer to 360 Myr. Discussions based on this moment of the geological scale are not outlandish. We remind the reader, however, that we left a "space", a looseness of around 12 million years in our argumentation that indeed favors the adaptationist point of view, which is the interval between the fossil record of †*Pederpes* and the beginning of the Carboniferous. All "geological distances" mentioned below as an argument can be increased by about 12 million years, which would favor even further each of our arguments, individually. We prefer, however, to "stretch the knot" against our argumentation, considering that effectively the first terrestrial vertebrate could be in fact older than †*Pederpes*, approaching the beginning of the Carboniferous, as was said above.

Minimum geological ages mean that the present knowledge of the first appearance of a character (apomorphy) is based directly on the fossil record. From these physical data (the fossil and its dating), a distance between its appearance and the appearance of terrestriality was established. From this distance the argumentation in favor of macroevolution was developed. However, new fossils may show that a character may appear even further back in time than indicated by our present evidence. This increases the geological distance to terrestriality, which corroborates our case in favor of macroevolution. Inversely, new fossils may show that terrestriality appeared before, diminishing the geological interval (distance), occasionally putting our argument in check. However, it should be noticed that the oldest known terrestrial fossil vertebrate is †*Pederpes*, which is circa 348 Myr old. In fact, we already increased the error margin by 12 Myr, up to the beginning of the Carboniferous, 360 Myr (an error margin that is, in reality, contrary to the argument in favor of macroevolution). Thus, the chance of these time intervals being reduced seems smaller than of their being increased.

Adaptationists apparently correlate geological and biological events of the Devonian with a progressive development of terrestriality. Clack (2012) summarizes many proposals in this regard, and presents a whole Chapter dedicated to the "Devonian world". We already presented our critical view of this stance in Chapter 2. However, synapomorphies of Osteichthyes indicated here (lungs and alternate movements of the paired fins) have been traced back in time, down to Silurian times, more than 440 Myr ago (Janvier 1996: 24, 181–182; *fig.* 1-23). The first terrestrial tetrapods are usually imagined as being from the Carboniferous, originating around 360 Myr (*e.g.* Clack 2012: 6, *fig.* 1-2). This leaves a lot of room (temporal span) from the origin of these characters and their use on firm land around 80 Myr. This time span is longer than the whole of the Cenozoic era, in fact representing too much time for us to consider microevolution as the main process directing evolution toward land. This issue must be considered in more detail and in depth. We deal with this subject below, dividing the issue into two distinct topics: characters through time, and the shape of characters at the time of their origins.

1 Characters

Characters and their discussion will be presented in chronological order. The sequence of characters in the cladogram corresponds with a sequence in geological time.

The chronological sequence overlaps characters, one being added to the others in organisms. Characters are added gradually toward the crown group, progressively uniting functional features. These functional characters appeared at different times in the phylogeny and in geologic terms, and therefore, independently of each other for terrestriality. The set becomes progressively more complete for life on land. But clearly, an increase in the adaptive speed of this set toward terrestriality, does not occur. This corroborates the proposal of macroevolution, instead of microevolution, as being the main evolutionary mechanism responsible for vertebrates having attained life on land.

For the reader to gain a better understanding of characters, and for easier reference, we must consider the remarks made in the previous Chapter. Following the approach therein, after the name of the character, we provide the number of the character as an Arabic numeral, between parentheses, followed by its function or effect, as a Roman numeral.

Lungs (1, III). Even when not considering the possibility that lungs appeared with the origin of the Gnathostomata, as some fossil specimens of †*Bothriolepis* suggest, some 445 Myr, the origin of these structures at 440 Myr is still geologically distant from the first terrestrial vertebrates, which appeared only about 80 million years later. This time span is longer than the duration of the Devonian or of the Cenozoic, each with a length of about 65 million years. Thus, to correlate lungs with terrestriality in Tetrapoda by means of natural selection, as an adaptation for terrestriality, becomes impossible to sustain, taking into account the geological time distance between one event and the other.

Alternate movements of the paired locomotor appendages (2, III). In the previous Chapter we considered this type of movement of the paired locomotor appendages as having evolved together with the origin of lungs. This character thus also appeared some 80 million years before terrestriality in vertebrates. The reasoning applied here is the same as for the origin of lungs in geological time.

Paired locomotor appendages (3–4, III). Sarcopterygii have paired muscular locomotor appendages, which is the feature that gives the name of the taxon. These were also provided with an elongate axial skeleton (forming the so-called paired, lobed fins). Each paired locomotor appendage has a single proximal skeletal element, which makes contact with the adjacent skeletal girdle. This combination of characters does not occur in other vertebrates and is considered an evolutionary novelty – an apomorphy – of the group. The characters involved in this complex appeared at the time of origin of the Sarcopterygii, or slightly later. Sarcopterygii originated at least 416 Myr ago (Zhu and Yu 2009). We may thus average the origin of these characters as being around 410 Myr. Because terrestriality in Tetrapoda appeared around 360 Myr, we have a geological interval of around 50 million years for natural selection to adapt these structures for their use on land. This represents a clear incongruence, because if a character is advantageous for its new function (whether or not it appeared specifically for this function), its refinement for terrestriality would have been immediate in geological terms (a few million years to become fixed in the common exclusive ancestral species of the remaining sarcopterygians and the closest derived species). These characters also form a functional column with which to elevate the body above the soil, for walking, which was possible also because the indispensable intrinsic muscles were present. The fact that there was a single skeletal element connecting the paired locomotor appendages (which we call limbs, in ourselves) with their respective girdles (scapular and pelvic) permitted a perfect rotation of each limb regarding the body, which is fundamental for walking. This rotating skeletal element was therefore ready for use on land. As well as forming a functional complex for walking on land, these characters appeared after the ability to breath air by means of lungs. The alternate movements of the paired locomotor appendages were already present, forming a functional set for terrestriality. Paradoxically, life on land would still take another 50 million years to become effective. This corroborates the idea that a process of adaptation for life on land should have been much quicker, geologically speaking, if they were "advantageous" for these animals. Fifty million years is equivalent to the duration of the Devonian period (*e.g.* Pough and Janis 2019: 96, *fig.* 6-1), or almost equivalent, according to other authors (63 million years in Pough *et al.* 2005: 198, *fig.* 9-1, for example). It becomes clear that these characters did not appear, nor became adapted as a function of adaptive scenarios for terrestriality. We stress, once again, that these structures originally had other functions, and they became coopted for terrestriality only a long time after their origin, and under special conditions, not related to selective pressures in the Devonian: the presence of other features for terrestriality, that could have exapted all together.

Choanae (5, III). According to Janvier (2004), these structures migrated independently to inside the mouth in Dipnoi and Tetrapodomorpha. They originated around 48 Myr before the tetrapods conquered land. Once again, we have a situation that is incongruent with the adaptationist view of the conquest of land, in terms of the geological time involved. The perspective of adaptation by natural selection requires advantage in the use of the structure, which would accelerate the process of fixation of the structure in the species. Such a refinement should be attained quickly in a few subsequent species belonging to its evolutionary lineage. Apparently Clack (2012: 388) agrees with our evaluation that the evolutionary process by adaptation must be brief, when she considers that the outer nares (on each side of the snout) enlarged quickly in Carboniferous forms. We stress that lungs were already present, as well as paired, columnar locomotor appendages, complete with intrinsic muscles. Even so, another 48 million years would still be necessary for terrestriality to occur.

The feet (the set of *carpus, metacarpus/tarsus, metatarsus and digits*), as well as the *bones radius, ulna/tibia and fibula* (10, 11, 111) are positioned approximately parallel to each other, and are articulated, respectively, with the carpus/tarsus, becoming the most appropriate paired locomotor appendages that ever emerged in vertebrate evolution for walking. They appeared in the fossil record about 395 Myr ago, and only began to be used on land around 360 Myr, that is, about 35 million years after their appearance. This is incongruent with the adaptationist view of evolution. If the character is adaptive for a certain function, it should be favored immediately, by means of some adaptation that would refine the structure for that function (in this case, terrestriality). The adaptationist thinking becomes even more questionable when one considers that structures that are complementary in function were already present in the record: lungs, alternate movement of paired locomotor appendages, which already had unique intrinsic muscles and a single proximal element forming the appendage, with a rounded proximal connection with the adjacent skeletal girdle, permitting their alternate movement. The anterior and posterior paired locomotor appendages, including feet and the parallel positioned skeletal elements for reinforcing limbs and sustaining feet (radius/ulna; tibia/fibula), were thus entirely present about 35 million years before their use on land. The character (mainly the set of characters already present) should have become adapted for land almost immediately (in geologic terms), for the adaptationist view to make sense. This represents one more fact which is incongruent with the adaptationist view of the origin of terrestriality.

The evaluation of Carroll (1988: 166), that requirements of locomotion and body support on land probably needed dramatic and rapid changes, support our analysis. For him, it is difficult to conjecture an adaptational process of long duration for an environment that is intermediate between that of aquatic sarcopterygians (that he called rhipidistians) and the terrestrial vertebrates (which he called simply tetrapods). In other words, the process of adaptation for the land environment had to be quick.

Loss of anal and dorsal fins (7, I, III). This character loss apparently occurred at the "†Panderichthyidae" split, about 395 Myr ago). This represents circa 35 million years for natural selection to refine this structure for use on land, which appears to be incongruent with the fact that, to be selected, the structure must be advantageous and the "adjustment" must be selective. It must thus be "immediate" in geological terms. Thirty-five million years is not what one would expect for the necessary rapidity that is usually associated with an adaptive selective advantage. This character is better interpreted macroevolutionarily, with regard to the conquest of land by vertebrates, particularly when considering that other characters of fundamental importance for the occupation of land were already available by vertebrates when this event occurred: lungs, columnar paired locomotor appendages with intrinsic musculature for alternate movement, feet, and choanae.

Functional neck (II, 12–13, 16, III). For a functional neck to appear, several changes are necessary. The last of these refers to the removal of the notochord from inside the cranium, which coincides with terrestriality in the cladogram. At least, no difference in geological timing between these events can be inferred. However, the loss of the series of opercular bones and of the extrascapular bone must have occurred at about the time of origin of the Tetrapoda, an event that must have occurred around 390–395 million years ago. This is 25–30 million years before terrestriality. Indeed, this loss may have occurred a little earlier because it is observable in the "†panderichthyan" †*Tiktaalik* (which was not included in our *fig.* 5-1). Again, the adaptationist perspective is that evolutionary advantages for use on land were the cause of the character having become fixed. But this time interval is incongruent with microevolution. This provides the opportunity for macroevolutionary explanations, as we have attempted in the previous Chapter. All the remaining characters necessary for terrestriality were already present: lungs, paired columnar locomotor appendages with intrinsic musculature permitting alternate movements, feet, and choanae.

Pelvic fins (8, III). The pelvic girdle is clearly developed at least in †*Tiktaalik*, a genus included herein in "†Panderichthyidae". The previous situation of the character is similar to that found in actinopterygians, including teleosts. In other words, the developed pelvic girdle is present at circa 384–382 Myr (the approximate age of the fossil remains of †*Tiktaalik*, according to Niedzwiedzki *et al.* (2010: 46, *fig.* 5). If we consider the first terrestrial vertebrates as having originated around 360 Myr, we would have about 20 million years for natural selection to shape this structure for use on land. This appears to be incongruent with the fact that, in order to be selected, the structure must be advantageous.

The selective "adjustment" has to be "immediate", in geological terms. And 20 million years is a time span that does not agree with the rapidity we normally correlate with the selection of an adaptive advantage. Furthermore, the structure was already present a long time before, with the pelvic fin already well developed and in an adequate position. Its size was already adequate, and the required refinements did not have to be very elaborate, considering that the articulation with the fin was already present and functional. We should also consider that the paired locomotor appendages in columnar shape were also present, with intrinsic muscles, and other structures were also present (lungs, feet, choanae), as explained above.

Rigidity of the vertebral columns (9, 15, 111). The zygapophyses appeared in the fossil record of sarcopterygians before the Poland fossil trails, that is, they date more than 35 Myr before terrestriality. Following our reasoning, this is too long a time for supporting a microevolutionary hypothesis of its origin for the conquest of land by vertebrates. The second step of development of zygapophyses regards developed zygapophyses being able to connect with each other (the anterior zygapophyses of one vertebra hinging with the posterior zygapophyses of the vertebra located immediately in front, and thus successively, from tail to head). Such connected zygapophyses are known from Sarcopterygii at least since the beginning of the Carboniferous. This practically coincides with terrestriality in terms of geological time. Only the use of the method of Blackburn (2002), in the previous Chapter, made the dissociation between these two events clear.

Depressiform head and body (6, 1). These characters are present in forms such as †*Panderichthys*, which has a maximum known age of around 385 Myr (*e.g.* Clack 2012: 60, *fig.* 3-2), 25 million years before the Carboniferous. Considering that †*Pederpes* (the first known terrestrial vertebrate) had a high cranium, one should be cautious about its use as a parameter for inferring micro- or macroevolution. In any case, the long geologic temporal distance from the origin of these characters and terrestriality indicates a relation of macroevolution (the characters would be considered to be exaptations for terrestriality), not microevolutionary events toward the occupation of land.

2 Abrupt Appearance of Several Characters

Several authors have conceived the origin of terrestriality as a consequence of environmental changes occurring through geological time. This view derives from the writings of Darwin himself, in which species originate by natural selection (a selection promoted by the environment). Environmental changes through geologic time would progressively promote organismal changes by the natural selection to the very changing environment. The most advantageous characters for a particular environment are selected and become fixed in the species. We do not question this evolutionary process. Nor do the proponents of exaptation question this principle, not even implicitly. On the contrary, they explicitly mention natural selection as the evolutionary process by excellence (Gould and Vrba 1982). Consequently, the reasoning is that new environmental changes through evolutionary time would have provided new opportunities for the selection of new characters. These characters would have appeared by natural variations in populations and species. As we have seen in the previous Chapter, there are other evolutionary processes that appear in parallel or jointly with natural selection. As we have just seen in the previous section of this Chapter, the distance in terms of geological time may also be used as an argument favoring macroevolution instead of microevolution, at least in the case of several characters.

On the other hand, we must still consider another aspect, linked to geological time, always on the basis of the fossils currently known: contrary to the belief of Darwin regarding the evolution of characters, these characters appeared abruptly. The characters are the lungs, the alternate movements of the paired locomotor appendages, a single proximal element in these appendages, their intrinsic musculature, a depressiform (dorso-ventrally compressed) body, and a head with eyes positioned dorsally, loss of dorsal and anal fins, pelvic girdle comparatively well developed, radius/ulna and tibia/fibula parallel and articulating with carpus and tarsus together with digits (feet), loss of opercular and extrascapular bones, ribs positioned ventrally. In the following, we will illustrate a particular character. In †*Tiktaalik* the series of opercular bones seems to be entirely missing, and the same seems to be the case with the extrascapular series (Clack 2012: 196). It is possible that all these bones have been lost simultaneously, or almost simultaneously, more or less abruptly in geological terms. This being the case, they could not have been progressively selected.

Had the characters been progressively developed, jointly with gradual variations along the Devonian period, then there would be an argument in favor of microevolution. But what is known is precisely the contrary, that is, the abrupt emergence of these features constitutes one more argument in favor of macroevolution. The excessive focus on the conditions of the Devonian must be reframed.

Furthermore, the focus on the changes in the Devonian is not corroborated when it is noted that several of these characters were already present in the Silurian: lungs, alternate movements of the paired locomotor appendages, a single proximal element in these appendages, as well as their intrinsic musculature. In other words, four out of 16 studied characters, or 25% of the total, occurred before the Devonian.

3 Evolutionary Space for Independent New Invasions of Land by Vertebrates after the Devonian

An outcome of the results from this Chapter and Chapter 5 is that there is no space for the idea that new terrestrial vertebrates do not evolve because the Devonian geologic and environmental conditions do not exist anymore and will not appear again.

There are several arguments against this idea, as follows:

1. *The first four of the 16 studied characters emerged before the Devonian* (*fig.* 5-1). Therefore, they cannot be considered to be consequences of selective forces for terrestriality occurring in the Devonian. Thus, it is not possible to state that the Devonian period developed the single possible evolutionary scenario(s) for the conquest of land by vertebrates.
2. *The abrupt emergence of the studied features*, according to the current knowledge of the fossil record. Had selective pressures been responsible for tinkering these characters, they would appear progressively in the fossil record.
3. *The progressive, but completely disconnected appearances of the studied characters along the cladogram, and consequently also through the geologic time* (*fig.* 5-1) is another argument against the supposed importance of the Devonian for the conquest of land by vertebrates. The characters would have to appear all together or in linked groups at given levels of the geologic scale or at given levels of the cladogram, together with specific eco-geologic conditions of the Devonian. This is not noticeable studying the cladogram/geologic scale.

   Further, geologists and paleontologists never specified the evolutionary operational Devonian conditions at a given emergence of an apomorphy for terrestriality. The characters appear independently of each other, through the cladogram/geologic column. In our simplified cladogram of *fig.* 5-1, it is very clear that there are several steps in the cladogram after the emergence of †*Acanthostega* and the emergence of developed zygapophyses, which was the last exaptation we identified before terrestriality. This set of steps in the cladogram constitutes a gap of evolutionary novelties that indeed is a wide "space" for selective pressures for land to have operated at the time, but there is no evidence for such occurrence. The "space" or gap itself is a counter-argument against the adaptationist view. This shows that there was no special condition for terrestriality at any given Devonian moment. Further, for the ones who believe that the question relies mainly on the appearance of the feet, then all or most of all characters for terrestriality should have emerged in the cladogram and in the geological column after the emergence of the feet, considering these are the catalyzers of evolution toward land. However, most of the characters we studied emerged much before the feet and several others emerged well after them. No Devonian specific conditions through its continuum are known to have existed that could be said to operate to form one or another of the characters we studied here.

4. *The emergence of terrestrial vertebrates only after a complete set of exaptable characters was present.* Other taxa outside Sarcopterygii do not have most of the 16 characters that collectively exapted for land. Actinopterygii share two of them: lungs and alternate movement of paired locomotor appendages in slow movement. A basal actinopterygian subgroup, namely Cladistia (as *Polypterus*, for instance), have an axial skeleton (but more than one skeletal basal element, as shown in *fig.* 5-13) in the pectoral fin, as well as intrinsic muscles in the fin. However, no actinopterygians have the following characters we studied here. Chondrichthyans do not have any of the exapted characters we studied here, except maybe the †xenacanthid chondrichthyans, but only regarding character 3 (*fig.* 5-14). The lack of these apomorphies – characters 1–5, 7–16 in *fig.* 5-1), necessary for a collective exaptation and for taking to the ground, is the real reason why no other vertebrate taxon passed from water to land.

5. *The Devonian eco-geological progression encompassed every life form.* However, only the crown group of vertebrates (possibly together with †Baphetidae) took off to land. As we presented in Chapter 4, had selective pressures been the cause of gaining ground by vertebrates, it should be expected that the terrestrial group of vertebrates was polyphyletic. Therefore, assigning progressively changing conditions to the Devonian, and therefore to selective pressures of the time as the cause of gaining ground in vertebrates, should be reconsidered.

CHAPTER 7

Absence of Correlation among Characters and Asynchrony in Their Origins as Additional Arguments

The method of Blackburn (2002), and the analysis based on geological time, utilize the same reference for comparison: the origin of terrestriality. The characters were all compared with this reference. The monophyly of the terrestrial vertebrates also uses the reference of terrestriality.

However, when this referential is subtracted, the characters may still be referenced, but now among each other (the "self-reference of characters"). This is a study in which we are pioneers again, as well as all the other arguments presented in Chapters 4, 5 and 6, as far as we know, where we used monophyly and geological time as arguments for macroevolution (Triques and Christoffersen 2009), and in favor of the method of Blackburn, 2002 (Triques and Christoffersen 2015).

The arguments based on a discussion of the "self-referent characters/ features" sprung from nowhere, when reading for the first time the manuscript of Chapter 6 in the Portuguese edition. We perceived that there is no connection among the characters, both in terms of origin on the cladogram, as in terms of geological time.

The adaptationist, microevolutionary view implies that an adaptive evolutionary scenario through geological time selects all characters for their new geological conditions at the same time. In the context of the process of selection, things could not be otherwise. At least a relative proximity in time is necessary for microevolutionary reasoning to make sense. Individually, each character does not adapt the organism sufficiently for a transition between one environment and another to occur. When a synchronic (or concatenated) adaptation is not observed among characters, macroevolution must be invoked to better explain the evolutionary process.

A differential, asynchronic timing in the origin of apomorphies linked to terrestriality evidences the functional and structural independence of characters among each other, and relative to terrestriality. They are also not restricted to the Devonian. These facts add to all the remaining arguments discussed up to this point (geological time of origin of each character until terrestriality; monophyly of terrestrial vertebrates; and the results of applying the method

 | DOI:10.1163/9789004684911_008

of Blackburn, 2002). All these arguments are coherent with each other, when evaluated simultaneously.

The absence of synchrony in the appearance of characters in the fossil record does not belong to the method of Blackburn (2002), and deserves an analysis of its own. Such an approach seems to be self-testable, because it is verified simultaneously in the cladogram and along the geological scale (*fig.* 5-1).

The asynchrony and, at the same time, the morpho-functional independence of each character (on the cladogram and through geological time) with each other, and in relation to terrestriality is, on its own account, an argument contrary to teleological chance acting in the origin of these same characters. Thus, there is no correlation with the idea of selective pressures acting in the environment in a given time span. Such an asynchrony cannot be explained in the light of adaptive pressures that would have selected these characters through time for terrestriality (or another function, in case we were dealing with another evolutionary event). Character asynchrony and independence among each other represents, therefore, another argument in favor of macroevolution.

Further, at the end of Chapter 6, we dealt with the fact that we have not found any published proposal for evolutionary operational Devonian conditions at a given emergence of an apomorphy for terrestriality. The asynchrony of the emergence of the several apomorphies or sets of apomorphies obliterates the possibility of such a proposal – the possibility of the existence of operational scenarios for the evolution of a given trait for terrestriality – exactly because of the asynchronies we are pointing out in this current Chapter. The momentary scenarios would have to select synchronically the full subset of characters, which we see to be disconnected in the cladogram and in the geologic column.

Looking at *fig.* 5-1, it may become easier to visualize the reasoning developed above. For example, the loss of the subopercular and opercular bones (character 12) occurs when the extrascapular is lost (character 13), but only six levels above (and this number would be much larger if the cladogram were complete), when the notochord (character 16) is disconnected from the cranium, permitting the emergence of a functional neck. Such a disconnection occurs by chance (macroevolution), because several levels of generality on the cladogram were necessary for another character to appear and enable the neck to become functional (and, it is worth stressing, to become factual). If selective pressures (microevolution) were the motor process of this evolutionary event, each character would be linked to each other, in the cladogram.

Another example refers to the paired locomotor appendages (or limbs, in tetrapods being popularly termed legs, with their respective feet). Characters

1 and 2 (lungs and alternate movement of the paired locomotor appendages, respectively) were conceived morpho-functionally as being connected, in Chapter 5. Thus, they form a morpho-functional unit. They appear at the same moment in the cladogram (*fig.* 5-1) and are close in geological time. Thus, several connections may be attributed to these characters (morpho-functional, phylogenetic, and synchronicity in geological time). As we have seen, the alternate movement of the paired locomotor appendages could have originated with the appearance of a gaseous vesicle (the lung, which becomes modified into a swim-bladder in teleosts). This could have been a response to a body density similar to that of the water. However, this set of characters does not have other known connections with the remaining characters studied herein, at least in terms of what is known by the cladogram and in terms of what is known of the time of origin of these characters, as they appear in the fossil record.

The alternate movement of these appendages initially appears disconnected, with the single skeletal element at their base (femur, humerus – the character 3), and with the intrinsic muscles (character 4). In our simplified cladogram (*fig.* 5-1), a single level separates the alternate movement of the appendages from the single proximal skeletal element (character 3) and from the intrinsic musculature (character 4). A complete cladogram would present some additional levels (*e.g.* Zhu and Yu 2009). On the other hand, the geological distance we were able to infer between both character sets was 24 million years. This extended geological time speaks against linking both character sets, as would be necessary from the point of view of explaining their origin by selective pressures (microevolution) for environmental adaptation toward land.

Six levels later (in this simplified cladogram) the feet appear (character 10). The geological time lapse we were able to infer after the emergence of characters 3 and 4 is approximately 20 million years. This disconnection between both events is indicative that they may have originated independently (whether in fact selective pressures ever existed, because these characters do not appear incomplete in the fossil record, in order to be later "shaped" by natural selection, as would be expected in the case selective pressures were involved in the origins of each character). Also, the temporal distance of circa 20 million years, as well as the distance of six steps of hierarchical levels, indicate a disconnection among these sets of events and among the environmental characters in which they were supposedly formed (if these conditions really were important for these processes).

The zygapophyses (character 9) appear eight levels before becoming functional as interlocking zygapophyses, the character 15. This number would be much larger in a more complete cladogram. This also strongly challenges microevolution as being the evolutionary process involved.

Finally, lungs (character 1) appeared around 30 million years, or a little more, before the choanae (character 5), as far as we have been able to infer. Together, they are used for breathing air in terrestrial vertebrates. As the choanae permit breathing without the need to open the mouth for taking air, the loss of internal body humidity, or simply water, during the inspiration and expiration of air, is diminished. Further, the simple fact that the mouth is closed for most of the time, diminishes the water loss by evaporation to the environment. Therefore, the combination of lungs and choanae is advantageous for a terrestrial vertebrate. However, there are three steps in our simplified cladogram (*fig.* 5-1) that disconnect both characters in evolution. Certainly, the sum of three hierarchical levels of distance, together with 30 million years in time span, is sufficient to frame macroevolution as the evolutionary process involved, because the large disconnection between the origins of these characters is evident.

This disconnection between characters and character sets, along the cladogram and through geological time, differs from Darwin's proposal of adaptive continuity of characters relative to environmental changes through geological time (and which represent the so-called adaptive scenarios). When the environment changes gradually, natural selection is also supposed to select gradually. What we find in the cladogram of *fig.* 5-1 is a disconnection of most characters involved with terrestiality, and are thus better explained as having evolved under a macroevolutionary perspective.

CHAPTER 8

Synergy among Arguments and Summary Considerations of the Main Arguments

We have presented several arguments in favor of macroevolution as representing the main evolutionary process responsible for the conquest of land by vertebrates. Macroevolution was redefined by us in Chapter 3 as a consequence of exaptation(s). Thus, our arguments are in favor of macroevolution for the passage of vertebrates from water to land. Our arguments in this case consequently are in favor of the predominance of exaptive events, against the current adaptationist view of evolution. Thus, we have relegated microevolution (selective evolutionary process) to a role of less importance, in this case for vertebrate evolution. The arguments are of quite distinct natures:

1. the monophyly of terrestrial vertebrates (the "crown group", that is, the taxonomic group that includes living land vertebrates and the remaining fossil vertebrates that roamed on land, form a group originating from a single exclusive ancestral species);
2. the exaptive nature of all analyzed characters, excepting one character necessary for the transition from the aquatic to the terrestrial environment. That is, the result of the application of the method of Blackburn (2002) to a cladogram that includes the crown group and the main subgroups of osteichthyans concerning the theme of the transition of vertebrates to land revealed that practically all 16 characters evaluated are exaptations, and only one character among them represents a possible adaptation;
3. the collective nature of exaptations, that is, it was necessary to have by chance the emergence, not only of several, but of all necessary features that then could exapt all together and at once toward land; therefore, the consequence is that the resulting taxon towards land is monophyletic, instead of paraphyletic;
4. geological time indicated that the majority of the evaluated characters appeared (already complete) dozens of millions of years before terrestriality, in a manner that is incongruent with a microevolutionary process and indicating that macroevolution is clearly much more relevant than microevolution for explaining the conquest of land. Complete aquatic characters that are considered to be advantageous for terrestriality would be refined for terrestriality immediately in geological terms (a few million years), because of its putative advantage for living on land;

 | DOI:10.1163/9789004684911_009

5. the abrupt emergence of the studied characters in the known fossil record, instead of a gradual emergence;
6. in the case where some taxa are taken out of the crown group and placed slightly more basally in Tetrapoda, the occurrence of aquatic forms closer to the †Baphetidae + the crown group is in itself a counter-argument against the adaptationist view; at the same time, characters previously considered in this book as exaptations for terrestriality continue as such, even under changing topologies of cladograms appearing in the literature;
7. the disconnected nature and emergence in the phylogeny of several characters and character sets that are independent of each other, which is contrary to the idea of continuity of natural selection as proposed by Darwin. Microevolutionary explanations require that these characters would emerge synchronically, or at least in concatenated form, in order to work together in an adaptive fashion toward land by means of natural selection;
8. absence of connection among the characters, both in terms of origin on the cladogram, as well as in terms of geological time. At least a relative proximity in geologic time is necessary for microevolutionary reasoning to make sense. Individually, each character does not adapt the organism sufficiently for a transition between one environment and another to occur. When a synchronic (or concatenated) adaptation is not observed among characters, macroevolution must be invoked to better explain the evolutionary process;

9A. logically, to be considered as adaptations for terrestriality, features 1–5, 7–9 in *fig.* 5-1 should have emerged together with feet (character 10 in *fig.* 5-1), because these traits do not work on land without functional feet for living on land. As there are steps in the cladogram between these features and feet, this is a new argument against adaptationism in the conquest of land by vertebrates, independent of the method of Blackburn (2002), and not correlated with other arguments listed previously and above. Further, the geologic time scale shows a gap in time between the emergence of these features and feet, corroborating the argument;

9B. as a corollary of 9A, considering that feet (character 10), as well as characters 11–14, are not functional on land without the presence of developed, interconnected zygapophyses (character 15), these features (characters 10–14) cannot be considered adaptations for land. This is because there are five to six hierarchical levels between these features and character 15, in our simplified cladogram of *fig.* 5-1. As the interconnecting zygapophyses are necessary to prevent the vertebral column to collapse due to the weight of the viscera while walking on land, the previous characters (in the cladogram) are useless on land. Therefore, they could not have been adapted for land in the absence of developed and interconnected zygapophyses. Indeed, characters 1–5, 7–9 also cannot be viewed as adaptations for land for the same reason;

10A. logically, features 1–5, 7–9 in *fig.* 5-1 cannot be viewed as preadaptations for terrestriality because, to be useful on land, and to receive selective pressures toward functioning on land, a foot useful for walking on land should already be present, which is not the case. Therefore, these features should be considered as exaptations for land – not preadaptations –, independently of the use of the method of Blackburn (2002);

10B. as a corollary of 10A, feet (character 10) and characters 11–14 cannot be considered as preadaptations for terrestriality in the absence of developed, interconnected zygapophyses (character 15), because they are not functional on land without a way to prevent a downward collapse of the vertebral column due to the weight of the viscera during walking on land. The five to six hierarchical levels of distance in the cladogram (*fig.* 5-1) prevent this hypothesis of preadaptation. The same is valid for characters 1–5, 7–9 (*fig.* 5-1);

11. characters are added gradually, in a cladogram such as that of *fig.* 5-1, toward the crown group, progressively uniting functional characters (which appeared at different times) for terrestriality. The set becomes progressively more complete for life on land. But clearly an increase in the adaptive speed of this set toward terrestriality does not occur. This corroborates the macroevolutionary view on the passage of vertebrates from water to land. One should consider that these features were ready when they appear in the fossil record or in living representatives (considering the soft tissue features), and the geologic time span considered in this book is that from the complete emergence of features to the possible origin of terrestriality, at the beginning of the Carboniferous;

12. the first explosive diversification ("adaptive radiation") of Actinopterygii occurred at the same moment as the emergence of the Tetrapoda, in the Frasnian. Both groups were under the same environmental conditions, but actinopterygians did not conquer land; this fact alone disclaims adaptation as being a dominant, universal, evolutionary process;

13. the second explosive diversification of Actinopterygii – the sister group of Sarcopterygii –, occurred at the very moment of the passage of tetrapods from water to land, that is, at the beginning of the Carboniferous. This represented the widest range of morphological diversification in the

actinopterygians. The first reasonably strong diversification of actinopterygian "fishes" occurred at the end of the Devonian, and faced the same selective pressures believed to have existed for aquatic tetrapods, prior to the conquest of land by these animals. Also, Actinopterygii continues to diversify through Permian times, when terrestrial vertebrates were diversifying. Therefore, several ecological niches were still open to new taxa;

14. ten to 25 basal actinopterygian subgroups were tinkered favoring future life on land, but this did not occur in any major chance event. All of them had lungs and enlarged skeletal elements in the pectoral fin, both of which could have been used on land;
15. further, Actinopterygians continue to diversify after the largest extinctions that occurred at the end of the Permian. At this moment in biological evolution, ecological niches might have been left available, as a consequence of that large extinction. Yet, actinopterygians did not leave terrestrial descendants, such as those of the crown group Tetrapoda.

The natures of these 15 arguments were discussed first independently of each other in this book. Thus, in our book we are not dealing with the unfolding of a single argument. Each argument has different reasoning origins and is based on different data sets. In other words, different means of study and evaluation reach the same result. Each and all arguments refute microevolution and corroborate macroevolution as being the fundamental evolutionary process that explains the passage from an aquatic to a terrestrial way of life. Therefore, we have a synergic situation, corroborating by mutual consent. It may be a matter of discussion whether the different natures of the arguments and/or the synergic situation that corroborates mutual consent constitute themselves new arguments in favor of exaptations and macroevolution (and against microevolution) in this case of vertebrate evolution. The arguments numbered as 1 and 3 above may be the most notorious combinations that mutually and synergistically explain and corroborate each other: the monophyly of the crown group Tetrapoda and the collective exaptation toward land. That is, if the process of evolutionary passage of vertebrates from water to land had not occurred through collective exaptation, the crown group would have to be polyphyletic.

From the vantage point of the distribution of characters (apomorphies) in the cladogram, still another type of synergy appears. Characters that originated before, in the cladogram, join characters that appear later in the cladogram, in such a way that functions or functionalities become added. Even so, an increase in the adaptive velocity of taxa in the direction of terrestriality is not seen. For example, from the alternate movements of the paired locomotor limbs up to the origin of feet, we have been able to infer 45 million years, along which time several features were added, namely lungs, femur/humerus, intrinsic muscles for the paired locomotor appendages, choanae, a clearly developed pelvic girdle, and the appearance of zygapophyses. Even so, more than 35 million years passed until terrestriality was attained. At this point, a much smaller number of apomorphies appeared (1/3 of the total studied), basically linked to the functional neck. This progressive superposition of useful characters for the terrestrial environment along the cladogram is not followed by a shortening of the geological time for attaining terrestriality. As we have seen in Chapter 6, after the tracks discovered in Poland, another 35 million years have passed until terrestriality was completed.

We envision another synergy, now between exaptation and the monophyly of the vertebrates, as an additional argument in favor of macroevolution. We have identified a mistake of many researchers in attributing exaptations to a continuous selective process, which would inexorably lead to the conquest of land. This reasoning is contrary to the idea of exaptations. Because it is thought that there are selective pressures for a certain environment or life style, these researchers think that "it is to be expected" that natural selection would lead to the origin of the necessary characters (for terrestriality, in our case), in one way or another. This mistake may be demonstrated by the fact that terrestrial vertebrates do not form a polyphyletic assemblage. This is the synergy between the monophyly argument reinforcing the argument of exaptations made clear by the method of Blackburn (2002). When characters are the result of environmental conditions (adaptive scenarios), these same conditions would have produced characters for the terrestrial environment – and, consequently, for terrestriality – in several distinct taxonomic groups, such as Actinopterygii, Chondrichthyes, and others, or even in more than one group of Sarcopterygii. As terrestrial vertebrates (as defined in this book) are not known outside of the "crown group" Tetrapoda presented in *fig.* 5-1, we may conclude that we have a corroboration of the idea that exaptation is a valid concept, and actually a necessary concept in evolutionary biology, as we have already argued at the end of Chapter 5 (Exaptations: Misunderstandings of the Concept).

The abrupt origin of several characters in the currently known fossil record provides another corroboration, or synergy, among ideas favorable for macroevolution, as we have discussed at the end of Chapter 6.

Thus, the adaptive scenarios linked to the Devonian must be carefully revised. Our results do not corroborate their relevance for the passage of vertebrates from water to land, indicating that they should be discarded from now on.

The method of Blackburn (2002) permitted us, with the characters available, to clearly distinguish exaptations from adaptations in the transition from the aquatic to the terrestrial environment, in the evolution of vertebrates. Consequently, the viability of the method is made clear and, at the same time, the importance of its use becomes evident for the comprehension of the evolutionary process. Gould and Vrba (1982: 13) predicted that secondary adaptations for a new function (effect) may occur after the exaptation of characters. These secondary adaptations become one of the effects of exaptations.

Our results corroborate the idea of Gould and Vrba (1982) that limbs represent an exaptation for terrestriality, and the proposal of Triques and Christoffersen (2009) that the conquest of land by Tetrapoda must have been mainly related to a chance event, attributable to exaptations. Clack (2002: 2) says that many evolutionary changes that led to the conquest of land occurred by chance – being in the right place, at the right time – rather than being the outcome of a directed process. In other words, 13 of the 15 anatomical characters studied (practically all characters, except a depressiform body, and the removal of the notochord from inside the cranium) and two behavioral characters (life in shallow water and alternate movements of the paired swimming appendages) were exapted for terrestriality, instead of adapted for this new life style. Thus, adaptationist hypotheses linked to terrestriality in vertebrates should be reconsidered. Subjective consistencies in isolated arguments should be avoided. These arguments should be linked to well corroborated cladistic hypotheses for their substantiation.

Arguments derived from adaptive scenarios (for example, increases or decreases in levels of O_2, CO_2, presence of terrestrial invertebrates, and other similar considerations) may be important from a microevolutionary perspective, but are not necessary in a macroevolutionary context. Adaptive scenarios may be tricky, because they only make sense in a microevolutionary context. They generate a tautology that erroneously diverts the understanding the researcher has on evolution. They force the researcher to always think in microevolutionary terms, when the focus is directed toward the mentioned scenarios. Clearly, the presence of terrestrial vegetation and arthropods, before vertebrates were able to progress on land, were necessary. However, these events occurred much earlier, since Silurian times. Thus, it is difficult to connect them causally to the evolutionary process affecting those vertebrates that made the transition from water to land.

The higher significance of the concept of exaptation, when applied to the transition of vertebrates from water to land, with the method of Blackburn (2002), is that permits us to conclude clearly and with discernment, that terrestriality in vertebrates probably did not occur as a main result of adaptations to terrestrial conditions (adaptive scenarios). Rather, terrestriality became possible after the independent origin of a set of apomorphies, each character arising through geological time, but not correlated with variations in aquatic or terrestrial environments. In other words, these apomorphies appeared by chance relative to the terrestrial environment. Finally, when this particular set of apomorphies became sufficient for the occupation of land, an important transition to a new environment became possible.

Further, our hypotheses of exaptations and macroevolution in the conquest of land by vertebrates are testable, while the previous adaptationist hypotheses (presented in our Chapter 2, but also in Clack 2002, 2012 among many others) are not. It is possible to test our hypotheses by means of new cladograms (systematic phylogenetic hypotheses) on the same taxa we are dealing with, from a higher level of universality (Gnathostomata) to a lower level (Tetrapodomorpha, or to a slightly lower level, including Tetrapoda, "†Panderichthyidae", and †Baphetidae, at least). Higher level phylogenies (that is, of wider universality) are necessary for testing the monophyly of terrestrial vertebrates. Phylogenies encompassing Osteichthyes and several of its subgroups are necessary to test exaptations by means of the method of Blackburn (2002). The depth of the phylogeny corresponds to a large diversity that encompasses all of the concerning features. Phylogenies on Tetrapodomorpha, or of a slightly less encompassing taxon, are necessary for testing exaptations in the case one or more taxa are taken out of the crown group (in competing hypotheses, when it is the case). The disconnected nature of features related to the passage from water to land needs a cladogram that encompasses every taxon that possesses them. The argument on the geological time gap existing between the origin of features is important for terrestriality in vertebrates. Terrestriality itself may be tested by means of more data on the matter. These data may maintain, diminish, or enlarge the gap, affecting the results on the original hypothesis. The abrupt emergence of the studied characters in the known fossil record may be tested by the discovery of more fossil information that could "fill in the blanks" regarding the hypothesis. We did not find literature tests for adaptationist hypotheses. However, when we proposed our own hypotheses (*e.g.* Triques and Christoffersen 2009, 2015, 2017, 2018 and herein) automatically we were testing the very adaptationist hypotheses as a whole, and we found support for chance events, that is, exaptations in the conquest of land by vertebrates, and therefore, for macroevolution instead of microevolution.

Some philosophers of Science consider only propositions that can be tested by experiments that can be replicated anywhere in the World to be scientific. Phylogenetic systematics is a line of thinking in biological systematics that permits exactly this replication of "experiments". That is, phylogenetic analyses themselves may be replicated, provided that the researchers have access to the material or literature, whenever it becomes needed or interesting for any biological purpose. Thus, testing phylogenetic hypotheses is part of the daily practice of biologists concerned with biological systematics (including paleontologists). Therefore, to use new hypotheses concerning the vertebrate passage from water to land is a direct corollary of researches on the theme. The testing of our hypotheses thus depends only on the new phylogenies that are frequently being published. We used the recent phylogenies of Clack *et al.* (2016) and Pardo *et al.* (2017) to test, in this book, our own previously presented proposals on exaptation, for instance. The results of these tests are reported at the end of Chapter 5, and they corroborate our original proposals. Indeed, after the mentioned tests, a new argument against adaptationism has emerged.

Gould and Vrba (1982) not only presented the new term exaptation, but also even proposed that "the order and arrangement of tetrapod limb bones is an exaptation for walking on land" (Gould and Vrba 1982: 11). At the time, no test was proposed for discerning adaptation from exaptation. Triques and Christoffersen (2009) used the monophyletic nature of the terrestrial vertebrates to suggest the occurrence of exaptations in the passage of vertebrates from water to land, as well as the geologic time span, and these arguments indeed represented two tests for the adaptationist proposals, that disabled them at once. This test can be done again and again. Indeed, the trackways of Poland described by Niedzwiedski *et al.* (2010) was a test that increased our argument on the time span. On the other hand, the proposals of paraphyly of †placoderms and †acanthodians add to our argument on the monophyly of terrestrial vertebrates, as we discussed in Chapter 4. The use of the method of Blackburn (2002) by Triques and Christoffersen (2015) was a test for our own previous proposal (Triques and Christoffersen 2009), and is discussed shortly thereafter.

On the other hand, the publication of the method of Blackburn (2002) brought a distinct way to discern exaptations from adaptations, and therefore, to test the matter. The implementation of this method (Triques and Christoffersen 2015) clearly disables the adaptationist proposals, feature by feature, and as a whole. Had a single feature that we dealt with not emerged in evolution, all of which were necessary for the collective exaptation toward land (Triques and Christoffersen 2015, 2017, 2018), implies that we human beings would never have existed, together with any other terrestrial vertebrate.

Therefore, for those adepts of the "testable" definition of Science, our proposals are all scientific, in this specific sense (and, therefore, in any other sense). The available tests for discerning exaptations from adaptations provided new support for macroevolution (exaptations) rather than for microevolution (in the sense of adaptationism), in the passage from water to land by vertebrates. Included in the available tests are not only the method of Blackburn (2002), but also the monophyly *versus* polyphyly (not envisioned before the proposal in Triques and Christoffersen 2009). The geological time gap (also used by Triques and Christoffersen 2009) added further support to the exaptive perspective. Several phylogenies proposed by adaptationist researchers themselves (apparently all paleontologists working on the theme) turned against the adaptationist proposals. We would like to highlight that we did not publish any phylogeny on the theme, but only used those already published by others. Therefore, it is not possible to attribute a bias against microevolution to any of the phylogenies we referred to, as we did not publish any of them.

It becomes clear throughout this book that a phylogenetic analysis is necessary and fundamental for the understanding of the evolutionary process that is involved, whenever the issues of adaptation or exaptation, as well as of micro- or macroevolution, are invoked. The evolutionary process can only be correctly understood under a cladistic framework . Without a cladogram, it is not possible to have an overview of character distributions through taxa. Therefore, in creating an adaptation hypothesis, an author must necessarily have at hand or create a hypothesis of phylogenetic relationships to support his or her hypothesis of adaptation.

Without a global understanding (overview) of the studied distribution of features, it is not possible to know precisely at which point in the evolution of an inclusive taxon each feature appeared. Without this knowledge, it is not possible to know whether the features appeared previously, together, or subsequently to the studied function (in the case of this book, terrestriality in vertebrates). Therefore, it is not possible to know whether the concerned process is an adaptation or an exaptation. Only the method of Blackburn (2002) can tell apart adaptations from exaptations, throughout an evolutionary lineage. However, this method requires a phylogenetic analysis, expressed by a cladogram. Blackburn (2002) states that it is necessary to have a stable, that is, well corroborated phylogeny. In our case, the derived position of terrestrial vertebrates, which appears in all available cladograms, is the key feature. Thus, in order to accomplish the analyses that permit the researcher to unravel most of the kinds of arguments we presented in this book, it is necessary to have at hand at least one well established phylogeny, summarized as follows.

The argument number one, at the beginning of this Chapter, on the monophyly of terrestrial vertebrates, needs necessarily a cladogram, because it is not possible to know what we are talking about in terms of monophyly without a phylogenetic analysis and its respective cladogram. The exaptive or adaptive nature of a feature (our argument number two, above), with the method of Blackburn (2002), could only be presented by means of a cladogram. Our argument number three, the collective exaptation for land of a number of features, can only be noticed by looking at a cladogram, as our *fig.* 5-1. All the reasoning on our argument number six, concerning which is the best interpretation (considering adaptations *versus* exaptations), if one or more taxon/taxa is/are taken out of the crown group, can only be inferred by looking at a cladogram. This depends on the accomplishment of a cladogram. In this case, it depends on which taxon or taxa are taken out of the crown group, which is/are its/their exact position outside of the crown group (as in Clack *et al.* 2016 and Pardo *et al.* 2017, for instance), so that the researcher can tell, from the new phylogeny, which is the correct new interpretation. Finally, the disconnection in the emergence of several features can only be discerned by means of a cladogram (our argument number seven). That is, only by looking at the cladogram it is possible to unravel this question, because the cladogram permits the researcher to have an overall perception of the group under consideration, by looking at it critically. At least in the case of the passage of vertebrates from water to land, it is necessary to have a general cladogram that comprises all the terminals which have the inherent features. We had to assemble one cladogram by ourselves in a simplified form, from previous papers published by specialists in the matter, in order to show the reader all these issues at once. Having a single figure at a glance is indispensable for the correct reasoning to appear in the reader's mind. Details on the matter of assembling a cladogram from others are presented in Chapter 5.

Only an encompassing and an at the same time simplified cladogram can give the reader an overall and simultaneous view of the involved taxa, of the distribution of characters, and of their functions and effects. Such a cladogram ensured our discernment for the evaluation of adaptation versus exaptation (argument number 2), for inferring the collective nature of exaptations (argument number 3), for understanding that the extraction of some terminals from the crown group not only did not change our previous reasoning, but also brings some new insights (argument number 6), in this specific case, of the passage of vertebrates from water to land, and finally for argument number 7, the disconnected nature of several characters and character sets among each other. Using partial phylogenies available in the literature would prevent us from seeing and, of course, understanding the facts indicated in *fig.* 5-1. Cladograms produced under different contexts may not show all taxa and characters of interest to our problem in the same figure, that is, in the same cladogram.

Further, studying only the phylogenies restricted to the universality level of the Tetrapoda would not permit us to unfold all pertinent characters. Our focus would fallaciously (although without bad intentions) become restricted to those features that would corroborate in an *ad hoc* fashion the adaptationist hypotheses.

Adaptationist hypotheses outnumber, by far, the alternative, exaptive hypotheses on the passage of vertebrates from water to land. This is because the arguments of Williams (1966), Gould and Vrba (1982), Triques and Christoffersen (2009, 2015, 2017 and 2018) were developed only after the method of phylogenetic systematics. They are being discussed (as in this book) and disseminated only quite recently. However, statements such as "the majority of researchers" only consider the adaptationist view of this case of evolution is not a scientific argument. "The majority" is not an argument in itself. We hope the reader will consider this, when approaching this book side by side with many others dealing with the subject.

Therefore, any kind of evaluation on the evolutionary process concerned cannot be done without a phylogenetic analysis, and consequently a cladogram, whenever the propositions on adaptation, microevolution, exaptation, or macroevolution, are stated (see Losos 2014: 5). The burden of proof mentioned by Williams (1966) currently implies the need for a well corroborated cladogram. After the publication of the basis of phylogenetic systematics – that Williams (1966) could not be aware of –, the new perception of phylogenetic relationships progressively became a sharp and necessary tool for the evolutionary thinking to become more precise.

Apomorpies that were exapted for terrestriality appeared by chance relative to the terrestrial environment. If some of them had not appeared, the transition to land would not have occurred. A good example of a fundamental exaptation for this transition, which might never have appeared, is the presence of a single proximal skeletal element in the paired locomotor appendages (femur/humerus), the adaptive advantage of which is still unclear.

Without the concept of exaptation, the reasoning becomes subject to the straight-jacket of adaptation, which fails to be justified, even among adaptationist evolutionary biologists. Such explanatory failures may now be convincingly attributed to the fact that the involved evolutionary process was not predominantly adaptive, but was mainly exaptive.

Thus, chance appears to be the key factor leading to the occupation of the land environment by vertebrates. This explains why only one monophyletic lineage of vertebrates succeeded to live in this environment, walking and foraging, after several dozens of millions of years of successive, but uncorrelated, origins of traits advantageous for live in water or, in some cases, of neutral adaptive value for this environment. But these characters, in their totality, and by chance, were of use and necessary for life in an entirely new environment, land.

CHAPTER 9

Exaptations, Not Preadaptations, at the Origin of Terrestriality in Vertebrates

One way to negate the importance or even the existence of the concept of exaptation is to invoke the concept of preadaptation. We argue here that exaptation is not preadaptation. Sometimes we hear from an interlocutor simply that preadaptation "already exists". No reasoning is presented together with this statement, as if it were clear enough in itself. Or as if everything were completely explained, at once, by this single and plain statement. In other words, the concept and word "exaptation" is simply useless. It seems to us that the concept or idea of preadaptation is the last bastion of resistance of adaptationists, so that this bulwark should be finally put down by us, in the case we are dealing with in this book. Indeed, other researchers already dealt with the question of preadaptation, showing that this concept is far from satisfactory. We will return to at least some of these authors, before presenting our own arguments against the use of preadaptation as an explanatory term and concept in the passage of vertebrates from water to land.

The textbook of Ridley (2004: 259) recognizes the criticism of Gould and Lewontin (1979), that we should accept a pluralistic view of evolution, rather than relying only on natural selection. However, in his single example for an alternative to natural selection, Ridley (2004: 259) presents to the reader a case of genetic drift for molecules. The author ends by stating that we need drift as well as selection in a full theory of evolution. For us, this is insufficient. Exaptation is or has the potential of being a leaping process in evolution, as was revealed in the preceding Chapters of this book. While Darwin saw natural selection as the main motor of evolution, exaptation permits leaps to occur in evolution. Furthermore, the collective exaptation of features may permit a taxon to shift its occupation from one environment to another. This represents a completely different evolutionary concept and perception, as we have demonstrated throughout this book. This concept may be called macroevolution. Macroevolution, when associated and redefined on the basis of exaptation, is a strong explanatory model that accounts for evolutionary processes in a way that natural selection alone cannot. Natural selection leaves gaps in our understanding of evolution, at least in some cases, as in the theme of this book. Because the conquest of land by vertebrates was not an adaptive process, all and every suggestion of adaptive scenarios has been demonstrated to

 | DOI:10.1163/9789004684911_010

be fallacious, by the very adaptationists working on them. As far as is currently known, genetic drift does not appear to be related to the evolutionary events responsible for macroevolution to occur. We are thus not going to prolong ourselves with this mechanism in this book. Therefore, to present undergraduate students, or the reader in general, only with genetic drift, as the single alternative process to natural selection, puts the reader/student/specialist's mind in a straitjacket. The book of Ridley (2004) aims to introduce evolution to undergraduate students, not to unravel all aspects of the evolutionary process. We are thus not blaming the author, nor the book, for not presenting a complete contribution to biology, of course. We only mention this detail as an example of what seems to be the usual view of evolution given to undergraduate students in biology. In fact, Losos (2014) is another example of a (somehow different) textbook (or guide, as in the title of the book) that little explores exaptation, although much is approached on macroevolution, in several distinct Chapters. Indeed, Gould and Lewontin (1979) provided a much more critical, slashing, and widely grounded view of the concept of natural selection. We understand that natural selection is an explanatory mechanism of the origin of species that is appealing and easy to understand by everyone (not only zoologists and botanists) around the World, and therefore is important as a weapon against religious fundamentalism. Therefore, it has to be highlighted whenever possible. On the other hand, speciation as a whole includes mechanisms other than natural selection: genetic drift, sexual selection, the genetics of differentiation and isolation, parapatric and sympatric speciation, hybridization and reinforcement, vicariance, migration, dispersal and colonization, chromosomal changes and maybe others (*e.g.* Freeman and Herron 2002: 403–435). Thus, the evolutionary process is more complex than natural selection and includes other mechanisms besides genetic drift. Microevolution seems to be regarded as more related to speciation and that does not explain certain evolutionary events that we are dealing with here. Concerning the theme of this book, exaptation and macroevolution represent the main process. Consequently, any objection that misleads the reader from considering exaptation and macroevolution should be rejected, with a definitive scheme of argumentation. This is the aim of this Chapter.

Ridley (2004: 260) says that the theory of adaptation of Darwin is "gradualist" (his quotation). For this author, Darwin's "gradualist" requirement is a fundamental property of evolutionary theory. We have shown throughout this book that this statement is incorrect, as the collective exaptation of several completely different and uncorrelated features is not a gradual process and clearly happened in the conquest of land by vertebrates. For Ridley (2004: 260), new adaptations evolve in small steps, starting with organs, behavioral patterns, cells, or preexisting molecules. Another way of saying this is to affirm that there is continuity among all forms of adaptation in nature. According to him, new adaptations evolve from preexisting adaptations in continuous stages, but this continuity assumes several forms. As an example of this continuity, Ridley mentions that the eye of vertebrates is an evolutionary novelty that evolved from ancestral photoreceptor cells located on the body surface.

Ridley (2004: 263–264), among many other authors, regards the continuity of all forms of adaptation "that we see in the World today" as a "contrast with, for example, a creationist view of life". His concern is, therefore, with a scientific, testable way of fighting religious fundamentalism. We understand this position. However, abrupt evolutionary processes do not validate any religious creationist statement. The concept of exaptation does not endorse creationism. Our new definition of macroevolution is based on the concept of exaptation and does not give any support for creationism. On the contrary, exaptation and our new macroevolutionary concept only explain clearly some evolutionary cases that are not correctly described/explained by progressive natural selection. Therefore, exaptation and macroevolution counter the last subterfuges of creationism.

The cumulative effect of many small changes may be such that something new appears (Ridley 2004: 264). Ridley uses the example of dinosaurs with feathers that, at the level of Aves, "add" the function of flight to these structures, which continue with their functions of display and thermoregulation. Ridley (2004: 264) says that the classical Darwinian term for a case such as this one is preadaptation. A preadaptation would represent a structure capable of originating a new function, such as a small structural change Ridley (2004: 264). He reaches and goes beyond the theme of this book, as we will see in his own words: a second example refers to the legs of tetrapods. Fish do not have legs, which appeared during the evolution of amphibians, and are now used for deambulating on land (Ridley 2004: 264). The fossil evidence, such as from †*Acanthostega*, suggests that, originally, the legs appeared to swim under the water (Ridley 2004: 264). The bone structure of the fins in one group of creatures became appropriate for deambulating on land, according to the view of that author (Ridley 2004: 264–265). On the contrary, we see that if the preadaptive view were true, terrestriality in vertebrates would have emerged several times and Neotetrapoda would be a polyphyletic group. Fish-like osteichthyans have lungs and alternate movements of paired locomotor appendages (and other features in our *fig.* 5-1) that are useful on land. Such features would compulsorily be considered preadaptations for land by authors such as Ridey, but would then lead to the mentioned polyphyly. Otherwise, these features should not be considered preadaptations at all.

Our view is quite different. We do not envision an adaptive sequence from the use in water to that on land, as is necessary for the concept of preadaptation to make sense, at least for the case approached in this book. The cladogram clearly shows that there was considerable distance among evolutionary levels since the appearance of characters (apomorphies) 1–5, 7–15 (of *fig.* 5-1) and the appearance of terrestriality. In this way the origin of a character is disconnected from the advent of terrestriality. Also, dozens of million years separate the origin of the foot from its second use, on land. If they had not appeared (for other functions, by means of other transformation series of characters), they would not exist for use on land. In the case evolution was simply addition of functions to already present and complete features, as Ridley (2004: 264) suggests, it would be continuous (or contiguous) in the cladogram and geologic column, but this is not the case. We perceive a rupture in the evolutionary process, explained by the concept of exaptation. On the other hand, lungs can hardly change their function, and practically do not add significant differentiations, at least as currently known, up to the origin of terrestriality, as defined herein. Even so, lungs are an exaptation for terrestriality, because they originated for life in water and not as part of a gradual colonization process for life on land. They are not the result of step-wise changes occurring during the evolutionary time that permitted (or led to) the conquest of the terrestrial environment. Lungs apparently provide, initially, new functional possibilities for animals living in the aquatic environment, but had as an effect the possibility of permitting life on land, several levels later in the cladogram. This is, exactly, exaptation. Adaptive gradualism is part of the concept of preadaptation, which is not observed either in the case of lungs, or much less in the cases of changes in function (in these cases, termed effects). As we saw in Chapter 6, geological time corroborates our point of view.

Symptomatic is the fact that the term exaptation is not mentioned in a textbook such as Ridley (2004), and little is explored in Losos (2014). Instead, the book of Ridley has a box (Ridley 2004: 265, *box* 10.1) to highlight molecular cooption, stating that the term cooption is often used to describe the evolutionary process in which a molecule takes on a new function, but with little change in structure. Also, in this *box* 10.1, it is stated that cooption is conceptually "much the same as" preadaptation. One term (cooption) happens to be used more regarding molecules, and the other (preadaptation) is primarily used when dealing with morphology (Ridley 2004: 265, *box* 10.1). Ridley (2004: 264) states that a preadaptation is a structure that happens to be able to evolve some new function, with little change in structure. Therefore, we find the same definition in his *box* 10.1, as for cooption. In our view, these statements of Ridley are misleading, because the cooption he defines in his box is abrupt in nature, and preadaptation is linked to the concept of adaptation. It is therefore progressive through evolution. Combining two different processes as synonyms prevents further discussion of the theme and provides educational closure. Exaptation is a clarifying term. If such a concept is not mentioned in a widespread textbook, it narrows the understanding of evolution for generations of students and people in general, exactly at a time of religious fundamentalism in the West.

In the case of limbs, the change in function (effect in Gould and Vrba 1982) represents a rupture in the evolutionary process. If feathers had not appeared to function previously on land, flying would not have occurred in Dinosauria or Aves. So, this represents another case of exaptation. Although Ridley (2004: 264) states that there was an addition on the functions of feathers (thermal isolation plus flight), for the feathers of flight, there was a change in use. These "additions" of functions, being different from the original function, are clearly exaptations for other uses, even though Riddley (2004) avoided naming them that way. The "added" functions are not preadaptations either, because there is no way to previously adapt – adapt in anticipation for the "additional" uses – (which is the literal meaning of the term, for the reader who first reads it) a structure for a completely different use. The feathers were exapted for flight, and a clear discontinuity can be defined in evolution. Further, without the initial emergence of feathers for thermal regulation or display, there would be no fight in avian dinosaurs. This is exaptation, not preadaptation. Feathers were never preadapted for flight. The extension of some of them, possibly for display, finally could be exapted for flight, but they were not somehow adapted for display and at the same time for flight. Even the conjecture of Ridley (2004: 263), that first the feathers were adapted for gliding, and then were adapted for flight, is misleading, because first some feathers had to appear, maybe by means of selective pressures for thermal regulation. Then, the feathers had to be exapted for display, being longer through a secondary adaptative selection, then they could be exapted for gliding, and this is exaptation, an abrupt change in use. This very view of Ridley (2004: 263) shows us that the idea of preadaptation is progressive instead of abrupt in nature. It is not wrong necessarily, but one has to argue in favor of it, and we are arguing against it in certain cases. At first, the concept of preadaptation does not necessarily invalidate or is not invalidated by the concept of exaptation. They may be both valid, but for different evolutionary situations. We are going to discuss them with our own arguments (Triques and Christoffersen 2018), but first we will review the arguments of previous authors on the theme.

Gould and Vrba (1982: 11) present a relevant question: what function do short feathers have for flying? Or, put in other ways: for what uses in terrestriality in

Tetrapoda are the following characters, (1) small gaseous vesicles for respiration, (2), only occasional (or irregular) alternate movements, (3) two proximal skeletal elements in limbs, (4) some small muscles in the paired locomotor appendages, (5) choanae that open outside of the mouth, (7) low profile dorsal and anal fins, (8) pelvic girdle only slightly more developed than previously indicated, (9) half-zygapophyses, (10) half a foot, (11) radius and ulna/tibia/fibula almost parallel to each other, (12) subopercle and opercle reduced, (13) extrascapular reduced, (14) ribs only half developed, and (15) notochord entering skull through a reduced foramen?

To qualify as preadaptations, they must be structures partly or progressively adapted for terrestrial life, but this is not what we see by examining characters (which do not present intermediate stages or progressively developed states for terrestriality through the fossil record) nor in their positions along the cladogram, relative to terrestriality nor are these partly developed features useful on land.

The concept of preadaptation has not been useful or didactic for the acceptance and comprehension of the concept of exaptation for most researchers in evolution. On the contrary, this view ended by enhancing the adaptationist view at the expense of macroevolution. At this point, we stress that exaptation and macroevolution were not terms created to substitute or to oppose concepts of preadaptation and microevolution, but they were introduced in order to complement evolutionary theory. They permit a more accurate perception of the evolutionary process as a whole.

Maybe one of the biggest merits in the conceptualization of exaptation and the publication of the term itself lies in that it permits us, at least in some cases, to note that evolution may be disruptive. The concept of preadaptation does not make this possibility clear, because the view of adaptation underlies this concept, inhibiting the perception of the role of chance in evolution. Not the chance in the variation of DNA, but chance in the appearance of characters adapted to certain functions that may be coopted for other very different, or totally new, effects.

The concept of preadaptation is gradualist, while that of exaptation is disruptive, or permits the view of disruption in the evolutionary process.

In the adaptationist view, the evolutionary process is always seen as being totally progressive and continuous. The concept of exaptation opens the door for new interpretations and perceptions in biological evolution, and maybe also in human evolution, in a historical and cultural sense.

Preadaptation is a complex issue. Several concepts and interpretations were apparently labeled with the term preadaptation. For instance, Simpson (1953: 190) considers that there were many examples of evolution labeled as preadaptation that could be put into nine categories listed by him. Simpson (1953: 188–189) states that the term "preadaptation" (quotation marks of the author) has been applied to a great variety of real or supposed evolutionary phenomena, from the appearance of a small mutation with selective value in the population in which it occurs, to the sudden appearance of a form that is monstrous in its parental population, but miraculously, one might almost say, adapted to some quite different way of life. A term used in so many different ways and, moreover, associated in many minds with the now usually rejected mutationist theory, is of rather dubious status. Many students have discarded the term entirely. Others, apparently still a majority, continue nevertheless to use it and have described under this name many evolutionary sequences of evident importance, especially in long-range evolution, and as regards higher [taxonomic] categories. The phenomena referred to are real, whether or not their interpretation is correct in any given case. The reader may call any, all, or none of them "preadaptation" as he/she chooses. Further, Simpson (1953: 189) proposes that most of these phenomena be called by the name, more neutral at present, of "prospective adaptation" (quotation marks of the author).

Simpson (1953: 189) noted that in order to realize the new functions of a changed environment, an organism must, at the moment when the change or the occupation of a new environment begins, have at least some functions prospective with regard to the new environmental functions. This statement of Simpson if far more advanced than Gould and Lewontin (1979) would expect from researchers that perceive evolution only as a result of natural selection generating adaptation. It is curious for us to notice that it may have had a turn toward a bipartite purism that did not exist some decades earlier than Gould and Lewontin (1979). Maybe the paleontologists working on the passage of vertebrates from water to land have always been purists (strict adaptationist), but not others. It is possible that this perception of Simpson (1953: 189), just cited above, is subtle enough to have gone unnoticed, even when also expressed by other evolutionary researchers of the time. He went further and, congruently with our research in this book and correlated papers, stated that probably all organisms, except for the most extremely narrowly specialized, and perhaps even those, do have characteristics that would permit some adaptive change (Simpson 1953: 189). In complete harmony with this last statement of Simpson, and complementing it, Gould and Vrba (1982: 13) state that the paths of evolution must be largely set by the size and nature of this pool of potential exaptations, and that exaptive possibilities define the "internal" (quotation marks of Gould and Vrba) contribution that organisms make to their own evolutionary future.

Simpson (1953: 189) states that preadaptation is practically universal. As prospective adaptation, it is to be recognized as a general rather than a special feature of the adaptive process. As we are going to discuss later in this Chapter, for us there is a clear-cut difference between preadaptation and exaptation. We can see from *fig.* 5-1 that, if we change "preadaptation" and "prospective adaptation" of Simpson for exaptation, then in the future it is very possible that exaptation will be shown to be a general feature of the evolutionary process, in parallel with the process of adaptation.

What called our attention very much was the following statement of Simpson (1953: 189–190): in the strictest sense, "preadaptation" (quotation marks of the author) involves the rise of characteristics, non-adaptive or unadaptive when they appear, but adaptive in some different, available environment. This is or seems to be the same concept of by-product or "spandrel" presented by Gould and Lewontin (1979), and that can be addressed directly to the concept of exaptation. Therefore, it is not unexpected that several authors depreciated the term and concept of exaptation in favor of the term and concept of preadaptation. We had to deal with this inconvenience for the sake of clearness, regarding the conquest of land by vertebrates, in a specific publication of ours (Triques and Christoffersen 2018). Misunderstanding in the use of the term arises mostly from a failure to define whether this or some other sort of prospective adaptation is meant (Simpson 1953: 190). Therefore, he highlights the fact that more than one concept may be included in the term preadaptation.

Consequently, Simpson (1953) evaluated a large number of examples labeled as preadaptation by some previous students and elaborated his own categorization system, with nine categories, in short, as follows (with some comments of ours).

1. A broadly adaptive character may, without change in itself, permit a change, or a narrowing in the realization of prospective functions. Simpson uses as an example a citation of Cuénot (1951) of the case of polyphagous species, which can change from one food to another, or the descendants of which can become oligophagous or monophagous. For us, this category is a simple finding that adaptations occur from similar structures or behaviors previously existing in related organisms. As far as we can see, it is a process of adaptation, step by step, progressive, not abrupt or disruptive. This case illustrates that, at the time, researchers were already aware that adaptation by natural selection is not the only process of biological evolution, and were trying to figure out what else was involved, with strong difficulty in depicting their own propositions. Today, indeed, the situation is not very different.
2. A variant arising in a population may be in the direction of an existing trend. Simpson evaluates that this is simply the role of mutations and variation in all trend evolution, *e.g.*, in increase of size among Oligocene

and Miocene horses, he says. We could copy and paste here our comments made for his category 1.

3. A variant arising in a population may be favored by selection in existing circumstances, and may initiate a new trend. According to Simpson, these are the so-called "key" mutations (quotation marks of the author), *e.g.* the mutation for claws in stylinodontine ancestors. This category seems for us simply to represent natural selection.
4. A specific adaptation in one ecological relationship may, without change, be requisite in another. Simpson mentions, as an example, adaptations of plants for dry soils or desert conditions, as preadaptations for an epiphytic habitat. As another example, Simpson cites lungs developed in aquatic vertebrates as representing a preadaptation for life on land. Exaptation apparently would belong to this category of Simpson. It is not our aim to sustain unconditionally the concept of exaptation. We are going to discuss this situation after category 9 of Simpson, below.
5. A specific adaptation in one situation may, by intensification, become adaptive to another situation more rigorously selective in the same direction (Simpson 1953: 191). For us, it is hard to discern this case from a proposition of simple adaptation.
6. An adaptive structural modification may, without essential change, serve a different function in a different adaptive relationship (Simpson 1953: 191). Simpson exemplifies this category stating that the broad type of digging foot may in some cases become used for swimming as in *Ornithorhynchus* and some moles, and the narrow type may be used for climbing, as in edentates. This category seems nebulous to us, as one feature may have more than one function. It is widely known that the climbing feet of tree sloths may be used for swimming. No preadaptive or exaptive nature is clearly devised, in our opinion.
7. An adaptive structure may be an essential continuation or intensification of a former morphological trend that comes to serve a different function (Simpson 1953: 191). He gives as an example the case of a seal (*Leptonychotes weddelli*), that uses its enlarged incisor to open holes in ice, while its relatives have smaller (less enlarged, in his words) incisors, and do not open holes in ice. That is, the enlarged nature of incisors gave rise (or opportunity) for even larger incisors to arise in evolution, possibly by increased natural selective forces, although Simpson does not put it in these words. This may be indeed adaptation from a previous condition, in a continuous selective process, although Simpson did not state this in a clear way. Our comments on category 6 may also be repeated here.

8. A structure adaptive in one way in an ancestral group may be transformed to serve quite a different function in a descendant group (Simpson 1953: 191). This is one of the abundant, striking features of evolution, and many examples are familiar to everyone: transformation of the reptile (dinosaurian, in current terms) walking limb to the bird's flying wing, or of the mammalian walking limb to the bat's flying wing, or to the cetacean's swimming paddle; transformation of a reptilian (amniote, in current terms) jaw articulating bones to mammalian auditory ossicles; transformation of a breathing lung to a hydrostatic bladder in fishes, and later, incorporation of the bladder in the auditory apparatus [in some groups, as ostariophysans, clupeomorphs and others]; and many others (Simpson 1953: 192). Noticeable is the following statement of Simpson (1953: 192), that this sort of change has seldom been called preadaptive. Comparing this with the statement of Ridley (2004: 264), that a preadaptation is a structure that happens to be able to evolve some new function with little change in structure, the difference may be restricted to the "transformed to serve" *versus* "with little change in structure". In about 50 years, the perception of students on the theme seems to have changed considerably, because by the 1950´s and earlier, "this sort of change has seldom been called preadaptive".

   Simpson (1953: 192) goes further, stating that his category 8 is not sharply distinguished from categories 6 and 7, usually called preadaptive and equally involving a change of function, or from category 3, sometimes called preadaptive, and equally involving a change in the direction of evolution. From these statements of Simpson, and from our comments on categories 3, 6 and 7 above, it becomes clear why our critics only state that "there is already" the concept of preadaptation, without further reasoning. For them, ideas and concepts are not clear, and their knowledge does not seem to be very strong.
9. A character nonadaptive or unadaptive in an ancestral group may be adaptive in a descendent group (Simpson 1953: 192). According to Simpson, this is preadaptation in the strictest sense of the word. We ask how a nonadaptive or even unadaptive character at its origin can be "preadaptive" (quotation marks are ours) for future use. For us, this cannot in any way represent a fact. As far as we can discern, this category of Simpson is what Gould and Lewontin (1979) present as a by-product or "spandrel" and can be termed an exaptation.

Categories 4 and 9 seem to be different cases of exaptation, but we prefer to continue using "exaptation", in detriment of categories 4 and 9 of Simpson for preadaptation. This is because category 4 of Simpson for preadaptation may lead us to conclude that natural selection is inexorably responsible for selection from one environment to the other. Plants would be selected until the epiphytic condition is reached, and vertebrates would be directed by selection in order to reach the terrestrial environment. This is not true, simply because terrestrial vertebrates form a monophyletic group. The concept of exaptation evidences the role of chance in evolution, when one environment or previous use of a character changes to another environment or use. Thus, the foot may have appeared as a refinement of the hydrofolium in basal sarcopterygians, but in fact another structure could have appeared, maybe for an even better fitness in the water, such as appears to be the case of the paired fins of †*Eusthenopteron*, for example (*figs.* 5-2 and 5-12). If feet had not emerged as a feature variation, they would not had been selected for better maneuvering in water, and the passage of vertebrates to land would never have occurred. Therefore, the evolution of vertebrates from water to land was not a simple case of natural selection toward land, or of emergence from water. (If so, terrestrial vertebrates would be a polyphyletic group, as we have already argued.) We indicated the occurrence of the collective exaptation of features in Chapter 5. The concept of preadaptation does not clarify this issue as well as the concept of exaptation, so we prefer to use this last concept and term. To refer to "category 4 of preadaptation of Simpson" (or to category 9 of preadaptation of Simpson, when it is the case of by-products) instead of to "exaptation", makes communication more difficult.

Category 6 of preadaptations of Simpson does not appear to be clearly distinct from category 4, but here structural adaptation is mentioned, instead of environmental adaptation. In any case, it was the naming and conceptualization of exaptation that permitted objective criticisms to adaptationism, and also led to the method of Blackburn (2002), corroborating our perception that this concept and term clarifies several issues in evolution.

We also prefer the term exaptation instead of preadaptation, because, as seen in Simpson (1953), there is a diversity of distinct situations that have been labeled as preadaptation, but which are not covered in more recent books (*e.g.* Freeman and Herron 2001; Ridley 2004; Losos 2014). The reader may gain a poor or false view on the subject: only two, maybe three cases of preadaptation developed by Simpson (1953) can be correlated with exaptation under that perspective.

Simpson (1953: 193) states that categorization of complex natural phenomena may not render clear-cut cases. Some examples of preadaptation could be placed in more than one category at the same time, according to him. On the other hand, the concept of exaptation may be clearly discerned from adaptation, when a well-supported phylogeny is available, by way of the method of

Blackburn (2002), as we saw in Chapter 5. Therefore, paraphrasing what some evolutionary spokesmen said to us, "there is already exaptation". Contrary to preadaptation, exaptation cannot be put into more than one category. If the term preadaptation includes several distinct concepts, as shown by Simpson (1953), confusion is due. With the term and concept of exaptation, mainly with the aid of a well corroborated phylogeny, the distinctions between exaptation and adaptation is clear. Thus, we do not need "preadaptation" anymore, for the sake of clarity.

A curious statement of Simpson (1953: 193) is that "no one but an extremist like Cuénot" (quotation marks are ours) has called 1 or 2 "preadaptation" (quotation marks of Simpson). We considered category 1 as part of an adaptive process, above, as we stated above for category 2. Simpson evaluates that this is simply the role of mutations and variation in all evolutionary trends. In short, Simpson (1953: 193) considers that Cuénot mistakenly interprets adaptation as preadaptation. We will see further in this Chapter why this is an important argument favoring exaptation instead of preadaptation in the conquest of land by vertebrates. According to Simpson (1953: 193), "Cuénot, indeed, went to lengths that may fairly be called ridiculous, as when he called the tendency for men to go in for one sport or another 'preadaptation des sportifs'", (Cuénot 1951, *ap.* Simpson 1953: 193). Further, Simpson (1953: 193) highlighted that extreme latitude in use of preadaptation while still implying or stating that it is something different from adaptation has, indeed, created a supposed problem that is semantic rather than real. We will take the opportunity to highlight that Cuénot, so drastically criticized by Simpson (1953: 193), is the very author of the term preadaptation, as we will see below.

Obviously, the very term preadaptation includes the term "adaptation". Therefore, "preadaptation" induces the student or the reader to believe there is some previous process of adaptation, because adaptation, besides being a noun (a term) also refers to a process of evolutionary biology. The student or the reader is persuaded that for a preadaptation to occur, a previous similar or nearly similar adaptation is necessary. The person is induced to believe that the preadaptation is an adaptation for an intermediary or similar condition, before the next step in evolution: an adaptation for the present environment, which becomes the focus of research. Regarding the theme of this book, a preadaptation would be imagined as representing an adaptation to an aquatic environment that is similar to the terrestrial land environment. Our findings reported in this book are exactly the opposite. Exaptations for terrestriality in vertebrates emerged dozens of millions of years before the terrestrialization occurred, and clearly for aquatic use, in aquatic environments disconnected from land. In other words, such aquatic environments are not necessarily located at the edge of the water. Thus, the correct term in the conquest of land by vertebrates is exaptation, not preadaptation.

Even category 4, exemplified by lungs as adaptations in water and then used on land, were also exemplified by vegetational adaptations for dry climates, and then used in the epiphytic habitat. The second case may lead the reader to suppose category 4 implies that similarities in the environments are the explanation for the "pre"-adaptative process. Therefore, more than one meaning may be included in this category of Simpson. In the case of exaptation, there is a single meaning. Simpson (1953: 194) explicitly admits that in categories 1 to 8, the essential features seem to be adaptive throughout. That is, the characteristics that were prospective adaptations with reference to the descendant populations arose adaptively in the ancestral population. In the exaptations we studied, a continuous process of adaptation (toward land) is not in accordance with our findings, considering the geologic time gap and the results of the method of Blackburn (2002). These showed several "steps" (hierarchical levels) from a feature's origin to its final use on land.

The recent general book on evolution by Losos (2014) shows "preadaptation" only in a few examples for preadaptive, instead of adaptive, features. But the concept itself is not mentioned or discussed.

Gould and Vrba (1982: 11) take the opportunity to discuss preadaptation. They aim to erect a solution for the problem of preadaptation. They evaluate that the concept of preadaptation has always been troubling for evolutionists. According to them, the concept of preadaptation was necessary as "the only" (quotation marks ours) Darwinian solution to an old taunt that incipient stages of useful structures could not function as well as the perfectly tinkered features (how functional is 5% of a wing?). It seems to us that the "taunt" (as Gould and Vrba said) seems to be a valid challenge for the adaptationist view of evolution. The answer Gould and Vrba (1982: 11) offer is clear and perspicacious: the incipient stages must have performed in a different way, not in the same way. They give the example of thermoregulation for feathers, not flight or any other kind of animal movement being correlated for this feature. Thus, Gould and Vrba argue that incipient stages must have had a different function.

Having clarified this first issue concerning "preadaptation", Gould and Vrba (1982: 11) go further, and state that the word preadaptation is wrong. If we divide the class of features contributing to fitness into adaptations and exaptations, they say, and if adaptations were constructed (by natural selection) and exaptations were coopted for their current use, then features working in one way cannot be preadaptations to a different and subsequent usage: this term makes no sense at all (Gould and Vrba 1982: 11). Therefore, exaptations are not preadaptations. The recognition of exaptations solves the conundrum.

Now we know (according to Gould and Vrba) that preadaptation is a term used incorrectly. Preadaptation is simply a category of exaptation considered before the fact. "Preadaptations" are only potential exaptations (Gould and Vrba 1982: 11). Thus, the term preadaptation should be given up in favor of another, also erected by Gould and Vrba (1982: 11), "preaptation", which is the potential, yet unrealized, exaptation.

Gould and Vrba (1982: 11) preceded us in stating that the order and arrangement of tetrapod limb bones is an exaptation for walking on land. They highlight that many modifications of shape and muscles are secondary adaptations for terrestrial life. Therefore, the concept of preadaptation is not necessary at the origin of terrestriality in vertebrates.

Casinos (2017) reviewed the concept of preadaptation and found a correlation of it with exaptation. Thus, the following question emerged: has something changed from Gould and Vrba (1982) to Casinos (2017)? Because of Casinos' (2017) conclusions regarding the recency of the paper of Gould and Vrba (1982), the issue of preadaptation had to be considered again here. Casinos (2017: 239) found that the origin of the term and the conceptualization of preadaptation occurred "as early as" 1901, by Lucien Cuénot (quotation marks are ours). We were not able to find the French words "préadaptatif" (preadaptive) or "préadaptation" (preadaptation) in the work of Cuénot (1901). Yet we have traced the term back at least to Cuénot (1909: 12) where the expression in French "caractéres préadaptatifs" (quotation marks ours), or preadaptive characters, appears.

Casinos (2017) attributed to Mayr (1976 *ap.* Casinos 2017) and Devillers and Guy (1996 *ap.* Casinos 2017) the recognition that the origin of the concept of preadaptation stems back to Cuénot. According to Casinos (2017), Cuénot coined the term. Casinos (2017) followed Stanley (1979 *ap.* Casinos 2017) in highlighting that Darwin (1872 [1901]) admitted the existence of features that would have arisen by natural selection for a given function, but that later could have served another function, with only slight modifications. Sepúlveda *et al.* (2011 *ap.* Casinos 2017), from this statement of Darwin, assumed a potential acceptance of preadaptive process as a direct consequence of an adaptationist point of view, according to Casinos (2017). Therefore, except if the term preadaptation is shown to have originated earlier than the researches of Cuénot, after Casinos' (2017) review, one must consider preadaptation as a creation of Cuénot and subjected to an adaptationist program. Thus, Cuénot's statements on preadaptation will be considered as part of the concept or the concept itself, from Casinos (2017) onward. As a result, the concept of preadaptation should be taken into account concerning the theme of the terrestrial origin in vertebrates.

The main conclusions of the review of Casinos (2017), on the connection he established between preadaptation and exaptation, are (1) that Lucien Cuénot introduced the concept and term preadaptation, in which the "emphasis" (quotation marks ours) was on the existence of structures associated with a particular function that could later serve another function; (2) that Gould and Vrba (1982) introduced the term exaptation as an alternative to the idea of preadaptation; and finally (3) Casinos (2017) asks whether the corresponding processes differ from adaptation. For us, the use by Casinos (2017) of the term "emphasis" denotes imprecision. As we are going to point out subsequently, Cuénot introduced imprecision to his concept of preadaptive characters throughout his publications, whereas Gould and Vrba (1982) presented a neat term and concept which they did not modify in their subsequent papers. Therefore, we consider that we can tell exactly what exaptation is, but preadaptation became (or always was) a fluid concept. We know exactly what we are talking about when we say "exaptation", but we do not know exactly what someone is meaning when they use "preadaptation", unless the person defines it when using the term. In itself, this is sufficient reason for dropping the term preadaptation, and for refusing the use of the concept of preadaptation at the origin of terrestriality in vertebrates.

Further, if Casinos (2017) himself makes such an issue of deciding whether preadaptation differs from adaptation, then the difference between preadaptation and exapatation is clear. On the other hand, this issue is not pertinent regarding exaptation, given the total and abrupt use of a preexisting feature. Further still, this difference between preadaptation and exaptation makes the importance of the term exaptation clear for the Science of form.

Simpson (1953) presented nine categories of preadaptation, some of them considered almost indistinct from others by Simpson himself (1953). However, some of these categories represent different concepts; some of them are completely different, others may be considered to be correlated. Therefore, preadaptation is not a concept anymore (in the sense of being a single concept) but a term that encompasses several, partial or totally different concepts. Exaptation, on the other hand, has a single definition and cannot be used in any other sense.

From Casinos (2017: 239), the extreme confusion created through the history of the term preadaptation becomes evident. We are using "term", because different concepts and interpretations were included in the term. For instance, Casinos (2017: 239) states that exaptation was proposed as a conceptual alternative to the idea of preadaptation. According to Casinos, Gould and Vrba (1982) considered preadaptation to have teleological connotations, "at least in its then current use", in the precise words of Casinos (2017: 239). If there was a "then" current use, it is clear that in different times the current use may have been different. This statement of Casinos makes it clear that more than one use may be correlated to the term preadaptation. Therefore, this is an argument for the complete discontinuity of its use in the scientific literature, for the sake of clarity. As a result, it should not be used when considering the origin of terrestriality in vertebrates.

Triques and Christoffersen (2018) aimed to use a cladistic framework to deal with the term preadaptation when considering the passage of vertebrates from water to land. However, these authors went further, and found some inconsistences in the association of preadaptation with exaptation. They considered them as different concepts (Triques and Christoffersen 2018).

Triques and Christoffersen (2018: 236) seized the opportunity presented by Casinos (2017: 240), who quoted and translated a statement attributed to Cuénot (1914): "preadaptations would arise in the closest environment ('le milieu le plus voisin', in the French original) to that finally occupied" (quotation marks ours). According to Triques and Christoffersen (2018: 236), this statement of Cuénot imprinted an idea of gradualism or continuism to the term preadaptation. In other words, the proponent of the term and concept of preadaptation himself included a perception of continuity into the term and concept he created. From this point on, the continuity of preadaptation is very close to the progressive view of Darwin for the process of adaptation by way of natural selection. That is why adaptationists wish to link the term preadaptation to the definition of exaptation. This happens to be a master play for turning exaptation into something that it is not: part of a continuous process. Even if the definitions of preadaptation and exaptation are shown to be the same at their independent origins, through time preadaptation became indelibly linked to gradualism by its own author. Thus, preadaptation became linked to the adaptationist view of evolution. That is another reason why exaptation is a term acutely necessary in the Science of form. It represents a clear counterpart to adaptationism.

The perspective of Cuénot that preadaptations would arise in the closest environment to that finally occupied indeed influenced other researchers. Consider the quotation Casinos made of Goldschmidt (1982[1940]: 102, *ap.* Casinos 2017: 241), that "preadaptive types of a population can migrate into a new environmental niche…" (quotation marks by Casinos). Obviously, Goldschmidt (*op. cit.*) was referring to specimens with variations of one or more features, which could use a new environmental niche. The statement carries the idea of progressively selected variations by the environment, which is simply adaptation by natural selection. Indeed, Casinos, just before quoting Goldschmidt (*op. cit.*), had already quoted Fisher and Stock's (1915, *ap.* Casinos

2017: 241) criticism, that if [regarding the concept of preadaptation], Cuénot "wishes us to understand that these preadaptations are fortuitous and accidental instances of the universal tendency to vary, then it only remains for us to say that he has added nothing to our knowledge" (quotation marks are ours). The idea of progression, of selected variations, is clear for us in the example of Goldschmidt (*op. cit.*), because preadaptation is correlated to feature variation by this author, at least in this case. It seems that Fisher and Stock (*op. cit.*) had the same opinion of the ideas of Cuénot on preadaptation: a simple process of adaptation. It seems clear for us that the statement of Cuénot that "preadaptations would arise in the closest environment" created the idea of a continuous process of adaptation. This permanently changed the idea or concept of preadaptation, becoming therefore the last barricade for adaptationism. That may be the reason why some modern textbooks exchanged Cuénot′s idea of continued processes of preadaptation for that of exaptation of Gould and Vrba (1982): uniting these two words in a single interpretation. Adaptationism was thus protected, by means of confusion and a conflation of concepts.

Because Gould never welcomed propositions of gradualism related to exaptation, it is not in fact possible to link exaptation to an adaptive, progressive process. Whenever an author considers exaptation to be a synonym of preadaptation, or exaptation as being part of a progressive process, Gould and Vrba (1982) can be evoked to show that this is not so. Even though the concept of preadaptation was proposed before exaptation, the concept of preadaptation was subverted by its very author. The idea was changed from an abrupt process to a continuous one. This is a key difference between both terms. Restoring the meaning of preadaptation to an abrupt occurrence in evolution is no longer possible. The notion of continuity was impregnated into the term by its own author (Cuénot 1914 *ap.* Casinos 2017: 240). This situation was clarified by Casinos (2017: 240), the most recent reviewer of the term and concept. Another key difference is that exaptation is a term and concept, whereas preadaptation is a term with several concepts linked to it (*e.g.* Simpson 1953: 190–192, Casinos 2017). One may say that preadaptation has several different definitions (*e.g.* Simpson 1953: 188–196) and one of them is or may be the same as for exaptation. Therefore, preadaptation is a misleading term in the scientific literature. For the sake of clarity, preadaptation is a term that should be dropped from the scientific literature.

Triques and Christoffersen (2018: 243) may also be evoked, because these authors disagreed with Casinos' (2017: 244) interpretation that non-adaptive by-products (of evolution of other features) that are later coopted for some use, represent a transitional situation from a condition of being neutral in the ancestor to become fully adaptive in the descendant. For Triques and

Christoffersen (2018: 243), this is a case of an abrupt process. Triques and Christoffersen (2018: 243) also argue that Gould and Vrba (1982), or Gould separately, never stated something similar to the above referred proposal of Casinos (2017). Therefore, again, exaptation is not preadaptation, and exaptation does not represent a continuous process.

Although Casinos (2017: 240) aimed to clarify the original concepts of preadaptation and exaptation, thereafter to see how the concept of preadaptation was dealt with in subsequent work, and finally to see if there really are differences between them, we would like to point out that there are no formal rules to consider as synonyms one concept that is prior to the other, as in taxonomic rules. That is, even if both terms might be regarded to be the same concept to begin with, having the first one of these shifted in concept and/or use, the second term that emerged does not have to be considered as a junior synonym, and dropped out from scientific use. There is no such formal rule in the Science of evolutionary biology. Thus, there are no normative proposals in Casinos (2017), or in any other similar paper. On the other hand, it is worthwhile to highlight the necessity and relevance of a paper like that of Casinos (2017). This paper analyzed and brought to light detailed information that is very useful and enlightening on the matter. Without Casinos (2017: 240), information on the perception of Cuénot that "preadaptation would arise in the closest environment", embedding continuity to the term – implicitly changing the concept from that point onward –, Triques and Christoffersen (2018) would not have developed this and other reasonings on the theme, for instance.

The term preadaptation means "somehow" adapted previously to a different environment or use. Beyond the progressive view of Cuénot on preadaptation, this perception of the feature being somehow previously adapted means that a process of adaptation has been in course before the new environment/use is attained. This perception does not exist for the concept and term exaptation. The publication of the term exaptation by Gould and Vrba (1982) permitted two different concepts in evolutionary biology to become available for evolutionary biologists. Therefore, this is another reason why the term exaptation is meaningful and needed to have been published.

Another, unpublished idea on the differences of preadaptation and exaptation is that preadaptation is considered by Cuénot to occur at the population level, while exaptation is a feature present in a species that later in phylogeny may be used by a descendent species (of the monophyletic group defined by the original species where the new feature is an apomorphy) in a different way and therefore is not related to population dynamics. According to Casinos (2017: 240), Cuénot (1941: 103 *ap.* Casinos 2017: 240) identified his idea of preadaptation with ecotype selection, which is an ecological subunit of an ecospecies, resulting from the genotypical response to a particular habitat. It was considered by Cuénot that every living and persisting being has by chance found the milieu that is suitable for it (Casinos 2017: 246). Therefore, it is possible to state that preadaptation as developed by Cuénot through several papers is the same as Darwin's adaptation; preadaptation is a synonym of adaptation, not exaptation. Casinos (2017: 240) highlights that the potential given by variation to part of the population would enable this part of the population to colonize an unoccupied niche (*sic* – sic of Casinos). In this way, these organisms would avoid competition with the rest of the population, according to Casinos. But we interpret this as a simple process of adaptation, not as any kind of preadaptation. What was presented in this paragraph indicates that preadaptation, as thought by Cuénot through his work, may be simply a detailing of the process of adaptation, but concerns the process for the passage from one environment to the other (or from one use to another), instead of adaptation to a changing environment, that was the original idea of Darwin. Indeed, the criticisms that preadaptation is simply adaptation was made by Fisher and Stock (1915 *ap.* Casinos 2017: 240), but with another reasoning.

Simpson himself (1953) followed the thinking of Cuénot, stating that preadaptation is used for an array of situations, from the appearance of a small mutation with selective value in the population in which it occurs, to the sudden appearance of a form that is "monstruous" in its parental population, but adapted to some quite different way of life. Therefore, "population" pops up several times in the considerations of Simpson (1953), and we see this situation as a consequence of the population correlation Cuénot imprinted to his concept of preadaptation. We consider the criticism of Fisher and Stock (1915 *ap.* Casinos 2017) to be correct. This is simply adaptation. For Triques and Christoffersen (2018: 237) preadaptation is, thus, a result of adaptive thinking, "as it could not have any other origin" (quotation marks ours). But for us, herein, preadaptation is merely a detailing of the process of adaptation, as we have said above. Simpson (1953: 189) found that many students discarded the term completely, and we are doing the same in this Chapter.

Therefore, Triques and Christoffersen (2018: 236) consider that the term preadaptation of Cuénot (1909), and also including preaptation of Gould and Vrba (1982), help to perpetuate the illusion that a feature was selected by a certain environment that is geographically close or that is similar to another environment. This view has already been exposed by Casinos (2017: 240). For him, this feature was "almost ready" (quotation marks of Triques and Christoffersen 2018).

Using a cladogram similar to that of our *fig.* 5-1, Triques and Christoffersen (2018) developed arguments in favor of exaptations, instead of preadaptations, for the conquest of land by vertebrates.

The first argument of Triques and Christoffersen (2018: 239) is that, looking at the cladogram (see *fig.* 5-1), it may be noted that all exaptations for terrestriality could be called "preadaptations" (quotation marks of the authors) for this lifestyle, but they only became effective as exaptations in †Baphetidae + the crown group Tetrapoda (= Neotetrapoda). That is, preadaptations become effective only in one (or few taxa, considering Clack *et al.* 2016 and Pardo *et al.* 2017, for instance) derived taxon (or taxa), being the remaining taxa – the vast majority of the preadaptation-carrier taxa – taxonomic groups that never turn preadaptations into adaptations for terrestrial life. If preadaptations do not become adaptations in derived taxa of the cladogram, then they are not preadaptations, indeed. Triques and Christoffersen (2018: 239) note that in the remaining groups in which the so called preadaptations occur, they do not become either adaptations or exaptations for terrestriality. The concept of preadaptation, they argue, becomes empty when no further change occurs through the evolution of a group (as seen by the *fig.* 5-1 in our theme). They must either end in a final adaptation or exaptation, as the case may be, in order to be evolutionarily meaningful. Therefore, preadaptation is a null concept.

Triques and Christoffersen (2018) were dealing strictly with the issue of the passage of vertebrates from water to land. We believe that whenever the apomorphies for a wider hierarchical level in a cladogram are said to be preadaptations for some, but not all of its subgroups, the concept of preadaptation is empty of evolutionary meaning. In cases like this, exaptation becomes the effective evolutionary event (Triques and Christoffersen 2018: 239), testable through the use of cladograms. The case of the conquest of land by vertebrates is emblematic in this sense.

In *fig.* 5-1 it is possible to see a large number of terminal taxa with "preadaptations" for terrestriality, in vertebrates that never developed further into terrestrial subgroups, as argued by Triques and Christoffersen (2018: 239): 14 sarcopterygian taxa, or 13, if we consider †*Pederpes* as being a terrestrial †Whatcheeriidae. Even Actinopterygii can be considered in this series (up to 15 terminals, therefore), taking into account that lungs and the alternating movement of paired appendages are apomorphies for Osteichthyes. If we include more known terminals belonging to Sarcopterygii, from which terrestrial groups could have originated, the number becomes further increased. Seven more taxa that could have originated terrestrial groups may be counted from Clack *et al.* (2016), for instance. The basal, non-teleostean actinopterygians, may increase the final number, as well.

Furthermore, Triques and Christoffersen (2018: 239) call attention to the fact that some basal sarcopterygians have living representatives: coelacanths and dipnoans, or lungfish, seen in *fig.* 5-1. Had "preadaptations" been capable of promoting rapid evolutionary changes, these taxa should not have remained only in their potential state, they argue. These authors further remind us that lungfish developed choanae independently from tetrapodomorphs (*fig.* 5-1), and still do not have terrestrial subgroups. We can remind the reader that lungfish (Dipnoi) emerged about 416 Myr, and all through this time, until the present, they did not produce terrestrial representatives.

Therefore, Triques and Christoffersen (2018: 241) found no connection of "preadaptation" (quotation marks of the authors) to known evolutionary processes. Thus, preadaptation could not be correlated even with adaptation. They conclude by stating that [the term] "preadaptations" (quotation marks theirs) induce readers to think that adaptations will necessarily occur through the forces of selective pressures, being only a matter of time for that, when this indeed was not the case (Triques and Christoffersen 2018: 241).

It becomes clear that the appeal of the term preadaptation is merely an adaptationist disguise, used when the *ad hoc* adaptationist hypotheses are not convincing anymore. Preadaptation is a term used in this way (otherwise exaptation would be the term used), in order to convince us that evolution is occurring strictly on the basis of adaptations (Triques and Christoffersen 2018: 239). Preadaptation is the final attempt for leaving no space whatsoever for alternative evolutionary interpretations (Triques and Christoffersen 2018: 239).

A criticism of the view that preadaptation identifies the process of modification of a character or trait, when two related sets of selective pressures are present (as seen in Shelley 1999, for instance), and its test by means of a cladogram, can be found in Triques and Christoffersen (2018: 241). In any case, the expression "two related sets of selective pressures" present at the same time, is an up-to-date way of stating what Cuénot (1914) referred to in his statement that "preadaptations would arise in the closest environment" (Casinos 2017: 240). It shows us that, even today, the idea of preadaptation is the same as for Cuénot (1914 *ap.* Casinos 2017: 240). Exaptation is a different concept, independent of the closest environment, or not dependent on two sets of related selective pressures, for a change in function. Again, preadaptation is not exaptation.

The second argument of Triques and Christoffersen (2018) is the collective nature of exapted features in gaining ground by vertebrates. That is, only the complete set of features necessary for living on land could be indeed effective, so that incomplete sets did not permit terrestrial occupation by vertebrates. That is why a single monophyletic group of vertebrates (Neotetrapoda, the crown group, or Neotetrapoda plus †Baphetidae, as in *fig.* 5-1) conquered land. The second argument is independent of the first (Triques and Christoffersen 2018: 239).

Several features correlated to terrestriality in vertebrates were already present and completely developed (rather than in progressive development) at several different hierarchical levels: Osteichthyes, Sarcopterygii and Tetrapoda (*fig.* 5-1), as noted by Triques and Christoffersen (2018: 240). These features accumulate through the cladogram (*fig.* 5-1), that is, along the evolution of the Osteichthyes. Only when a set of features necessary for terrestriality is complete, may a collective event of exaptation occur, concomitantly, toward land, rather than one at a time. If only preadaptations were enough for the conquest of land by vertebrates, the previous existence of a complete set of characters would not have been necessary for a collective exaptation to occur. In this case, argue Triques and Christoffersen (2018: 240), several distinct taxonomic groups would have originated terrestrial vertebrates, rather than merely a single one, as shown by Triques and Christoffersen (2009). This is a case of synergy of arguments (collective exaptations on the one hand, and the monophyly of terrestrial vertebrates, on the other), one explaining the other more clearly.

Even after the emergence of the feet, delimiting the Tetrapoda, this lineage did not start the emergence by natural selection of several independent terrestrial groups (Triques and Christoffersen 2009), although Tetrapoda already had 10 "preadaptations" for land that were also living in the "closest environment", as advocated by Cuénot (life in shallow waters, the lifestyle I in *fig.* 5-1). The so-called preadaptations for land did not preadapt any taxa for land, as becomes evident. Only when more features necessary for living on land emerged, still for use in water, all of them exapted, collectively, toward land.

The third argument is the cumulative nature of so-called preadaptations, seen through the cladogram (Triques and Christoffersen 2018: 240). A conspicuous issue, verifiable in the phylogeny of *fig.* 5-1, is the gradual accumulation of "preadaptations", that occur along the phylogeny (Triques and Christoffersen 2018: 240). As we noted in Chapter 5, the following features were already present when the feet appeared: lungs, alternate movement of the paired locomotor appendages, that got intrinsic muscles and a single proximal skeletal element, choanae, the loss of the dorsal and anal fins, clearly developed pelvic girdle, occupation of shallow waters, and presence of zygapophyses. Even with the progressive accumulation of features that could be called preadaptations by adaptationist researchers, the progressively less inclusive hierarchical levels did not lead to terrestrial vertebrates (except for the crown groups plus †Baphetidae). Had there been a process of progressive natural selection leading to progressive adaptation for land through Devonian times, these progressively accumulated "preadaptations" should have given rise, independently, to different lineages of terrestrial osteichthyans. Not even this large set of "preadaptations" was enough, together with the totally formed feet, to produce several independent terrestrial groups of tetrapods (Triques and Christoffersen 2018: 240). Thus, the concept of preadaptation becomes devoid of meaning, at least considering the conquest of land by vertebrates. As seen in the previous argument (number two), only a complete (appropriate) set of features for land permitted the collective exaptation of them toward land, in an evolutionary event that we may denominate macroevolution (Triques and Christoffersen 2018: 240–241).

As an extensive and cumulative set of "preadaptations" may be seen in *fig.* 5-1 (as apomorphies in the cladogram), and considering that only the full set of these apomorphies exapted, in a collective mode, then it is clear that these "preadaptations" do not exist individually as such (Triques and Christoffersen 2018: 240). Each of these apomorphies, on their own, do not represent preadaptations for land, because only the full set of features necessary for living on land exapted, together, as we have said several times in this book. Therefore, the concept of preadaptation is not applicable to this case of animal evolution. If only the collective group of features could work on land, it is impossible to address the status of adaptation or preadaptation for each one of them, individually and separately. Therefore, a preadaptive or adaptive nature for the conquest of land cannot be attributed to these features. Only one collective exaptation for life on land occurred a single time in the evolutionary history of the vertebrates.

These features/apomorphies (*fig.* 5-1) were not "preadaptations" (quotation marks of Triques and Christoffersen 2018: 240) for anything, but the entire set enabled a collective exaptation for terrestriality. In short, the individual potentialities of the supposed preadaptations were therefore shown to be null (Triques and Christoffersen 2018: 240).

Further considerations on preadaptation, preaptation (of Gould and Vrba 1982) and plesiomophy are developed in Triques and Christoffersen (2018: 240).

The three arguments of Triques and Christoffersen (2018), contrary to the invocation of preadaptation against exaptation in the conquest of land by vertebrates, have a cladistic basis, that is, they can only be viewed through the plotting of apomorphies in a cladogram. Maybe the best sequence of arguments would be that preadaptations present in all of the basal actinopterygians and sarcopterygians seen throughout the basal tetrapods do not develop into adaptations for terrestriality (the first argument), not even with the accumulation of preadaptations along the cladogram (the third argument), and that only a complete set of apomorphies necessary for land finally led to a single origin of terrestriality by means of a collective exaptation or macroevolutionary event (the second argument). The latter represents a chance event, that is, the process of gaining ground by vertebrates is not preadaptive, nor adaptive, but exaptive. The first and second arguments could have been developed from the monophyly of terrestrial vertebrates, as an argument contrary to adaptationism, as developed by Triques and Christoffersen (2009), but the third argument (the cumulative nature of the so-called preadaptations in the cladogram) is only visible when observing an inclusive cladogram, with the relevant apomorphies plotted in it (as in *fig.* 5-1). From the collective argumentation developed throughout this book, it becomes clear that, whenever a hypothesis of adaptation is claimed, a well-supported cladogram should be studied in order to test it. This is part of the burden of proof Williams (1966) refers to as a hypothesis of adaptation. But we paraphrase him also concerning preadaptation hypotheses. That is, whenever "preadaptation" in invoked, a relatively stable cladogram provides a good test for it. However, Blackburn (2002) states that even when features emerge in the cladogram together with their functions, the tests of Andrews *et al.* (2002) should be implemented. We did not implement these tests in this book or in our previous works on the theme, because this was not deemed necessary, for now.

Paleontologists working on the theme focus their evolutionary analysis on the groups most closely related to the terrestrial clade (the crown group Tetrapoda). This restricted focus enhances the supposition of preadaptation regarding the conquest of terrestriality in vertebrates. Vorobyeva (2008: 124) evokes the term preadaptation (features that look like preadaptations, in the original) regarding the transition toward land in vertebrates. However, with a broader perspective, that becomes viable with a cladogram such as that in *fig.* 5-1, phylogenetic depth is introduced into the analysis, as shown by Triques and Christoffersen (2015, 2018). Therefore, not only a cladogram is necessary, but a broad cladogram that goes beyond the number of terminals is required (as shown in Triques and Christoffersen 2018: 244), and which is also inclusive in terms of hierarchical levels. So, in *fig.* 5-1, we approached Osteichthyes, but mainly Sarcopterygii, and inside this, mainly Rhipidistia, toward the terrestrial groups. A large number of terminals, that even so do not permit an overall view of taxa and features (such as in Clack *et al.*, 2016 and Pardo *et al.* 2017), will not make the real extension of these components clear. Therefore, the method of Blackburn (2002) will be used imperfectly (as already explained in Chapter 5 and in Triques and Christoffersen 2015). If not for these preconditions, the analysis concerning preadaptation made by Triques and Christoffersen (2018) would not have been possible.

Further discussion on the three arguments may be found in Triques and Christoffersen (2008: 241–244). References on cladograms as powerful tools to work on concepts such as adaptation, exaptation and preadaptation, including cladistic and empirical or other types of tests for pradaptation, can be found in Triques and Christoffersen (2018).

Looking at *fig.* 5-1, it is noticeable that after the emergence of lungs in an aquatic animal, several hierarchical levels were necessary for these features to be used for respiration on land. In the simplified cladogram of *fig.* 5-1, 14 hierarchical levels may be counted from one state (use in water) to the other (use on land). Therefore, this feature was not adapted for the aquatic environment and at the same time for terrestriality (Triques and Christoffersen 2018: 242). In terms of geological time, the reasoning is the same: about 80 Myr of disconnection between use by an aquatic animal and use by a terrestrial animal (Triques and Christoffersen 2018: 242). The disconnection is shown by a cladogram, but was already clear from a geological time scale (Triques and Christoffersen 2018: 242). One line of reasoning corroborates the other, in a synergy of different sources of information (Triques and Christoffersen 2018: 242). Therefore, no preadaptation can be demonstrated in this analysis of the case (the lungs), but rather a cooptation, that is, an event of exaptation (Triques and Christoffersen 2018: 242). This case shows the power of phylogenetic analyses and cladograms for enlightened reasoning, with no detriment to the usefulness of the geologic scale. Thus, cladograms may shed light on the evolutionary process of concern, generating arguments in favor of one and against another hypothesis. The geologic scale may be embedded in the cladogram, such as in *fig.* 5-1. The same reasoning used for lungs can be used for each of the exaptations studied by Triques and Christoffersen (2015) and here (*fig.* 5-1). Due adjustments in the cladogram need to be made for decreasing geologic time and for placing events that become gradually more subordinate in hierarchical groups, according to Triques and Christoffersen (2018: 242).

After this reasoning, and looking at *fig.* 5-1, one can see that preadaptation only "exists" (quotation marks are ours) as a function of an adaptation or an exaptation, that emerges later in the cladogram. However, this later adaptation or exaptation was not found by us in most of the cladogram terminals (indeed, in none but one). An evolutionary later change may or may not occur in taxa (terminals in the cladogram) that have these structures (Triques and Christoffersen 2018: 243). Thus, it is possible to state that preadaptation does not exist on its own (Triques and Christoffersen 2018: 242). Preadaptations can only be recognized after their future effectiveness in evolution (Triques and Christoffersen 2018: 242). This is the same as recognizing the concept of exaptation (Triques and Christoffersen 2018: 242). Preadaptation is once again seen to represent a nonexistent concept in the evolutionary process (Triques and Christoffersen 2018: 243). A term-and-concept that does not represent an evolutionary process and that does not necessarily lead either to an adaptation or to an exaptation, does not exist (Triques and Christoffersen 2018: 243). In simple terms, preadaptations that did not come to perform a further use are not preadaptations (Triques and Christoffersen 2018: 243). In addition, when they do, they are exaptations (Triques and Christoffersen 2018: 243).

Only the adaptationist reasoning benefits from the term preadaptation, because it surreptitiously leads the reader to believe that evolution can only occur by means of adaptive natural selection (Triques and Christoffersen 2018: 243).

The second argument of Triques and Christoffersen 2018) against the concept of preadaptation, linked to the origin of terrestriality in vertebrates, is that of the collective nature of exapted features. This argument describes a disruptive evolutionary process from life in water to use on land (Triques and Christoffersen 2018: 243). Otherwise, it would not be collective – that is, occurring at once –, in the cladogram. Had the event not been collective, several independent origins would have been found toward land in vertebrates or in sarcopterygians, at least (the monophyly argument of Triques and Christoffersen 2009). The collective nature is an argument in itself, but the conjunction of it with that of the monophyly of the terrestrial origin in vertebrates has a synergic convergence that strengthens the argument as a whole.

Triques and Christoffersen (2018: 243) argue that these features can only be viewed as having worked in combination, and that if a single one of these features had not emerged for use in water, the whole set of features could not have exapted for land, and terrestriality would never have occurred in vertebrates.

The view of evolution by means of natural selection, feature-by-feature, was already criticized by Gould and Lewontin (1979), to whom the organism evolves as a whole, and that parts of it must evolve somehow together. The collective exaptation of features toward land in vertebrates is congruent with their criticism. This finding of Triques and Christoffersen (2015, explained in Chapter 5) is exactly that features were exapted together to conquer land in vertebrates, that it was necessary to have a complete set of features for this exaptive process to occur, and that this was a fortuitous case of evolution. This finding explains, at least in part, what Gould and Lewontin (1979) were proposing in their paper. Trying to understand the conquest of land by vertebrates, feature by feature, is counter-productive: it is not possible to understand correctly an evolutionary process that depends on several unrelated features for another completely different use (terrestrial in comparison to aquatic). Particularly for preadaptations, the collective exaptation of features demonstrates that there were no parallel selective forces for living in water, on the one hand, and toward land, on the other, in each of the unrelated features necessary for land. The plotting of the studied features that are subject to adaptationist (and "preadaptationist", we would say) hypotheses in a comprehensive cladogram may show whether a preadaptive, adaptive or exaptive process occurred. The evolutionary process may be depicted from (or at least discussed by means of) the topology of the cladogram.

The findings included in this book, related to a cladistic framework, can be tested, by means of new phylogenetic analyses, in the sense of Hennig (1966). However, the single cladistic possible refutation for these results and conclusions (including the conclusion on preadaptation[s]), would be if a polyphyletic crown group of terrestrial vertebrates starts to emerge from future alternative cladograms of Osteichthyes, or any wider taxon (Triques and Christoffersen 2018: 244).

The conclusions of Triques and Christoffersen (2018) are impacting, as follows. Their results indicate that preadaptation is a null concept, with no real implication in the evolutionary process (Triques and Christoffersen 2018: 245). The concept or idea of preadaptation only benefits the adaptationist view of evolution (Triques and Christoffersen 2018: 245). The adaptationist view of evolution is a teleologic, environmentally directed view of evolution (Triques and Christoffersen 2018: 245) and therefore is an interesting argument or point of view only for environmental evolutionists (Triques and Christoffersen 2018: 245). Ecology and geological variations through time become the single important features of evolution, shaping the DNA variation, under this view of biological evolution. Morpho-physiologic and behavioral viewpoints for understanding organismal evolution is unmerited by adaptationism (Triques and Christoffersen 2018: 245). For us, these conclusions of Triques and Christoffersen (2018) may be applied to all situations we explored through this book.

We understand that the features themselves, have not only a real, but a fundamental role in biological evolution, that goes much beyond the mere process of DNA variation shaped by the environment: features tinkered by one selective force may be useful for a completely different use later in evolution, in a less inclusive hierarchical level of the monophyletic group for which the feature is an apomorphy. As Gould and Lewontin (1979) already demonstrated, this reasoning is also valid for features that emerged without a function, as a by-product of the evolution of other features.

The concepts of exaptation and by-products ("spandrels") give credit to zoology itself, in the study of evolution, not only to DNA variation, and to environmental selective forces. Organismal features have an importance that environmental evolutionary researchers antagonize, resist in accepting. This book gives credit and importance to zoology for the study of evolution, depicting in detail how to study, to show features, and to illustrate the importance of zoology *per se*. (This reasoning equally applies to botany, a discipline that likewise has merits of its own.)

This book also gives priority to the human brain, rather than to a technological approach to Science, or to new findings of specimens/taxa. The papers published "yesterday" are not necessarily better than the "older" ones, including Hennigian phylogenies among them. Thinkers through history of evolutionary biology brought ideas forth that we could unite and develop ourselves in order to present our own interpretations. No technology or new findings were necessary. The idea that papers or books published more than five years ago are outdated is wrong.

We dare to believe that this whole book will indicate to other researchers in zoology and botany the way (or one way, at least) to approach biological evolution and depict which evolutionary process is indeed predominant: this will not necessarily be DNA variation tinkered by environmental selective forces. General guidelines for this research, involving proposals for preadaptation, are provided in Triques and Christoffersen (2018: 245).

At this point in our book, we would like to stress that the arguments presented herein are not due to an accumulation of facts, to the recency of published data, to ultramodern techniques applied to fossils (such as computerized tomography, synchrotron, etc.), new fossil discoveries, or the most recent or high impact publications. We ourselves have never collected fossils professionally, nor do we study them directly, and neither have we produced ourselves phylogenies of the groups involved. Even phylogenies more than two decades old, such as those of Clouthier and Ahlberg (1996), that support the paper of Triques and Christoffersen (2015), and that are the basis for our Chapter 5, are highly informative and sufficiently detailed for our discussion. The phylogeny just published yesterday is not necessarily better or more elucidative. Thus, the idea that the accumulation of information, its recency, the use of sophisticated technologies, etc., are the basis for advances in Science, must be questioned. We use basically the human brain and the logic pertinent to our theme. As we have never had in our hands a single vertebrate fossil related to our theme, then the idea that only specialists that have access to collections may provide opinions on the subject at hand is an attitude we are forced to disagree with. Terms such as exaptation were published more than one generation ago, and highly pertinent criticisms have been provided even longer ago. The concept of exaptation was published long before a term was created to name it. Initially, we believed that the concept was published in the 1960s, but in the present Chapter we can find the same concept published at least in the 1950s. "Old" concepts, ideas, and terms prove to be more recent and pertinent than ever. Human knowledge cannot remain restricted by preconceptions such as those related to time, sophisticated technologies, or to the accumulation of data. But it is necessary that scientists give a little time to think about their data and use the available systems of knowledge, in order to arrive at sensible conclusions. Concepts published a long time ago may be quite illuminating and in need of revival. The human mind is capable of producing much more than the latest technologies, because it produces consciences. Yet, if the brain is not used critically, it will be worth less than simple new data.

CHAPTER 10

Science and Its Limits

Science is a construction of the human mind. There remains no other possibility. A new perception of reality opens possibilities for even newer interpretations. Clearly, the result of Darwin's convincing work on natural selection as the main engine of biological evolution had several consequences. One was the emergence of new arguments about how biological classifications would represent the outcome of the evolutionary process. Hennig's phylogenetic systematics was one of these new proposals. However, knowledge on phylogenetic relationships can be used as feedback on discussions on the evolutionary process itself, as seen, for example, in this book. Therefore, scientific paradigms generate new paradigms, and all of them influence each other. For some people (including researchers), a world of new possibilities will arise, but for others, their world will seem to be crumbling. In day by day academic life, the reader can imagine what may happen to a researcher's mind if his/her arguments for obtaining financial support for research are challenged by the emergence of new paradigms. This researcher, with his/her graduate students, and also the institutions that house them, would all possibly react in some way. The general public does not necessarily perceive what is at stake, nor are developments on these matters properly informed to the general public. One strategy of self-defence (for adaptationism) was described by Gould and Lewontin (1979): to avoid mentioning alternative reasoning or contrary facts.

1 A Statement on Scientific Paradigms

Most of the new insights we have gained on the theme of the conquest of land by the vertebrates were obtained under the Hennig paradigm of phylogenetic systematics. Indeed, the first argument obvious for us came from a Hennigian statement regarding the monophyly of terrestrial vertebrates. Observing a simplified cladogram in textbooks for teaching a course on vertebrates at the undergraduate level, the new argument emerged, 20 years ago. As we developed our careers under the Hennigian paradigm, it was natural for us to see biological evolution through this viewpoint: all other arguments (cladistic or non-cladistic) started to emerge from this moment on.

System thinking in biological systematics was only introduced a little over a couple of generations ago, by Hennig. Despite his revolutionary thinking,

 | DOI:10.1163/9789004684911_011

many original phylogenetic concepts have remained anti-intuitive for many biologists trained under the dominant Neodarwinian paradigm. Many evolutionary biologists, particularly those working with Genetics and Molecular Biology, and including most Palentologists, continue to think within the dominant framework provided by the Evolutionary Synthesis, that preceded Hennig's paradigmatic shift in world view. Unfortunately, the architects of the Evolutionary Synthesis themselves have represented one of the main obstacles for the dissemination of Hennig's views. Paleontologist George Gaylord Simpson vowed that he would never cite Hennig's works in print while alive. Geneticist Theodozius Dobhzansky and botanist George Ledyard Stebbins simply ignored Hennig's views altogether. Ernst Mayr was the only one to criticize Hennig in print (Mayr 1974), provoking the only rebuttal in English by Hennig himself (Hennig 1975). Despite Hennig's masterful argumentation, this was not sufficient to shatter most evolutionary researchers' faith in the dominant Neodarwinian (the modern synthetic theory of evolution or New Synthesis) worldview. As already pointed out in this book, Clack (2012) recognizes the New Synthesis as the origin or foundation for discussions on Devonian scenarios. It seems that paleontologists working on the theme became permanently linked to this paradigm. Further, Gould and Lewontin (1979) indicated that in the United States and United Kingdom, the undergradute education in the field of Biology only teaches the adaptationist view of evolution. This means that a different perception of evolution will be viewed with suspicion by the vast majority of biologists in the World, even today.

2 A View on the Sociology of Science

Science, like religion, represents no more than a set of institutions and rituals. And what are the scientists? A scientist is simply an honorific term (such as Professor, for example). Scientists are no more than people trained to do some useful activities in particular places, being committed to seeing the world in a particular way, just as any other professional. It is in this sense that scientists should be respected (Fuller 1998). But it is also notorious that skilled scientists frequently fail when applying their logic to other particular fields. Ironically, traditional Science, like religion, is exclusivist, absolutist, authoritarian, and coercive, producing a technological elite, instead of focusing on the democratization of knowledge. The frequent use of the passive voice in scientific texts tends to obscure this personification of Science, this detention of knowledge by a very few (Oyama 1991). The more than 300-year-old tradition of deleting all human action lying behind scientific representations has, inadvertently, given Science an almost divine authority, which is yet another detail approaching it to religious texts (Keller 1993). There is a considerable difference between what scientists promise or sustain in theory and what they do in daily practice, whether in the lab or in the field. Within Science, increasing specialization removes incentives to unify distinct branches of knowledge. From outside Science, there is the continuous threat of cuts in investment budgets, when maintaining material investment is necessary for the continuity of research. There are simply too many interests involved to be able to admit that a radical critic may be right (Barnett and Fuller 1998). Market goals are restricting the frontiers of research to private interests, relegating the role of universities more and more to simply providing courses and lectures. The western ideal goal of interacting teaching with research is being followed less and less by public universities. Finally, we may cite the myth of Science progress. Historical treatises on the Sciences usually picture Science as always progressing, inexorably, in the direction of truth. Many people are led to believe in an amazing and impressive progress of knowledge over time. Yet the preservation of this additional myth of Science may represent only a literary artifact, or a rhetorical appeal, in an age when the necessity for permanent public financing remains for the continuity of its research activities (Fuller 1998).

Nothing is more distant from the image of the modern researcher than that of a disinterested individual, focused only on the search for truth, that of an independent person totally compromised with his ideals and subject to the impartial judgement of those peers that have gained the Olympic heights of their incomparable knowledge, a knowledge which has become totally inaccessible to the population in general (Merton 1938).

Science in the XVIIth century, with Francis Bacon, was conducted inductively. In the XXth century, the hypothetic-deductive method became the fad and persisted up to this date. But a more adequate model for present Science, and long overdue, in our opinion, is based on the semiotic views of Charles Sanders Peirce (Cohen 1923). Peirce's abduction represents the process of analogical reasoning. This process is important because it justifies the genesis of new ideas in Science. Perhaps the moment has arrived to enhance the most important process in Science, the creative process, which permits one to view old realities in new ways. Instead of the normal and cyclical Science of yesterday, we seek a Science that is permanently revolutionary, that does not lose its transformational energy by becoming crystalized onto sectarian paradigms. The Science of tomorrow should consist of permanently creative dynamic movements (Fuller 1999). Another quality of the new Science that we envision is that of being constructivist-interactionist-systemic (Oyama 2001). Emphasis must be placed on interactive processes acting at several scales. Form and function should be seen as emerging into processes. An integrative Science must act on processes and products of development. When we focus on the system as a whole, we do not find discrete causes and effects, but recurrent dynamical processes (Oyama 1994). Develomental systems are organized hierarchically. Interactionism requires two-way changes of information (Oyama 1999). Nature, the subject of natural Sciences, is not a fixed essence, but an organism-environment complex moving and changing through time and space (Oyama 1993).

Today there are thousands of institutionalized scientists, sociologists, and humanists that view the search of knowledge simply as one profession among others. This mass of scientific workers teaches in classrooms, publishes their academic papers, or continue to develop their research, circumstantially linked to the most immediate market interests. Through their publications, that represent the currency to attain prestige and political power in Science, they compete for the priority of their ideas, for the acceptance of these ideas by their peers, and for the general public judgement concerning the utility and relevance of the produced knowledge. But Science as Faith is exerted by few. The true religion we call Science will always exist, as long as there are scientists that believe that it is worthwhile to fight for their opinions and personal convictions, even when they clash with the collective cultural episteme (that represents the socially constructed, ephemerous, present "reality"). The critical stance in Science and its success as the alter-ego of society depend on these permanently revolutionary movements that necessarily question the *status quo*. Thus, Science does not become exclusively reduced to relations of power and competition. Science also participates as an element of scientific rationality, which does not represent the only form of possible rationality, but will contribute to project a special light, among other possible views, about this particular world of ours. A world that we, stubbornly, obstinately, and by all means at our disposal, try to describe, interpret, and comprehend (Christoffersen 2001).

3 Epilogue

We believe the above three subsets of this Chapter explain why so many researchers resist abandoning the adaptationist program. It also accounts for the amount of rejection our novel ideas have encountered from the established scientific community, when our initial papers on the theme were first submitted for publications in the standard journals dealing with evolutionary biology and paleontology.

On the bright side, we hope our new conceptualizations of a classical theme will do much to reinvigorate the merits of the transformational system view introduced by Hennig into Evolutionary Biology. We also envision that this book, with about 100 thousand words, elaborated into reasoned arguments along the last decade, will do much to advance the case for the still incomplete Hennigian revolution. After well over a millennium of Essentialism, a time span of a little more than a single generation has clearly not been sufficient for System Thinking to dominate Evolutionary Biology and to become the standard paradigmatic outlook. The challenge of a definitive paradigm shift in Evolutionary Biology lies with the next generations. We are certain they will eventually succeed in completing the ongoing revolution.

Glossary

Acanthodians extinct group of vertebrates that may be defined by the presence of anterior spines in all fins, except the caudal. Their monophyly has been challenged.

ACCTRAN Acronym for accelerated transformation series. When it is equally parsimonious to distribute a feature in a cladogram in different ways, the researcher is allowed to choose whether he/she considers that the feature emerged first in the cladogram and reverted secondarily or the opposite. In the first case it is said to be an "accelerated" emergence of the feature, in a character transformation series where the character is reverted subsequently in the cladogram. The opposite situation is said to be a "delayed" emergence of a feature in the cladogram, and the corresponding acronym is DELTRAN.

Actinopterygians in skeletal terms, vertebrates with only one dorsal fin amongst those with bony fin rays (the osteichthyans); dorsal and anal fins structured only with fin rays (that is, the skeletal support for these fins – included inside the body axial musculature – does not become prolonged into the fin itself).

Adaptation in evolutionary biology, a feature is an adaptation only if it originated by natural selection for the function it now performs.

Adaptive convergence: similarity generated through time by selective pressures on features of unrelated taxa.

Adaptive scenario environmental landscape in which taxa suffer selective pressures generating adaptations through generations.

***Ad hoc* hypothesis** an argument or hypothesis formulated with the sole aim of legitimating or supporting a theory, not resulting from an objective and exempt understanding of reality; we believe that this is not necessarily a conscious task on the part of the researcher.

Amniotes tetrapods that have an amniotic egg as part of their life cycle. This "egg" is characterized by specific membranes, as the amnion, chorion and allantois. Most mammals do not produce an "egg" in the strict sense, but are popularly known for having amniotic fluid.

Amphibians the understanding of the monophyly in the sense of phylogenetic systematics of Amphibia is dubious because amphibians, as originally proposed by Linnaeus, contain only forms related to living animals. Such a group is currently named Lissamphibia, that may represent a monophyletic group. When fossil taxa were included at the end of 19th century, amphibians became what we would consider a paraphyletic group. The first or basal tetrapods (which were aquatic forms and sometimes are called Devonian tetrapods, such as the famous *Ichthyostega*), were also accomodated within "amphibians". Amphibians were

 | DOI:10.1163/9789004684911_012

yet to be considered a group to include the closest representatives to living forms (lissamphibians), and therefore amphibians are not a monophyletic group.

Anal fin see median fins.

Analogy features or feature states not belonging to the same evolutionary character transformation series.

Apomorphic state of being of an apomorphy.

Apomorphy an evolutionary novelty, or innovation. It can be a modification of a preexisting feature or a completely new feature. Apomorphies are the only valid kind of features to ground monophyletic groups (also named clades) in Phylogenetic Systematics or Cladistics. Gains, losses and/or reversals may be considered apomorphies, through a phylogenetic analysis, and are considered evolutionarily equivalent in this kind of analysis.

Autapomorphy an apomorphy of a single terminal in the analysis.

Basal taxon (or taxa) or [taxonomic] group(s) in a phylogeny, the taxon or taxa that split early. Roughly, this is similar to the stem group definition, except that the more "derived" taxa in the phylogeny are not necessarily the crown group. This definition does not imply that this taxon (or these taxa) are primitive in any sense. On the contrary, it has or they have their own autapomorphies and, therefore, has (have) evolutionary innovations of their own. See also stem groups.

Batrachomorphs part of the crown group Tetrapoda, and that include living "amphibians" (lissamphibians).

Bony fishes called osteichthyans, characterized by the presence of bony fin rays (whereas the other gnathostomes have keratinized fin rays). Composed by actinopterygians and sarcoperygians (the lobe-finned vertebrates).

Branch in a branched diagram, a link between a taxonomic group and another or other groups. These kinds of linkages express the known phylogenetic relationships amongst the studied taxa.

Branchial relative to the respiratory organ of basal vertebrates (and also its invertebrate closest relatives). Relative to the gills. From Greek, *branchia*, gills. See branchial arch.

Branchial arch skeletal elements of the branchial apparatus, or gills, where the gill filaments are inserted. The visceral arch.

By-product a feature that emerged in a taxon as a consequence of development by natural selection (or of genetic drift) from another feature. It is expected that by-products have no function in several or most cases. If later in the evolution of the taxon (that is, in a subgroup of the taxon) the useless by-product becomes used somehow, it is said that the by-product was exapted for the new use and that it now has an effect. Function-free, duplicated genes may be by-products of gene mechanisms, but may assume new functions (effects), which would be a case of molecular macroevolution.

Carpus the set of small bones that compose the wrist.

Character in Biology, an attribute of biological organisms. Also called a feature or a trait. Structure is a term often utilized as a synonym of a character. Some authors consider a trait to represent a complex of characters, but others consider them as synonyms. In this book, we are usually referring to an apomorphy.

Character state the state of being of a character or feature, at every moment of a transformation series during the biological evolution of this character. Sometimes, character and character state are used as synonyms, as for a hair. However, modifications observed in the hair during its evolution are to be considered as character states through time.

Choanae paired internal openings on the roof of the mouth, or palate, for the entrance of air into the lungs. The internal nostrils.

Clade the same as monophyletic group in Phylogenetic Systematics or Cladistics.

Cladistics often considered a synonym of Phylogenetic Systematics. In this case, clade is a synonym of monophyletic group and Cladistics is the name of the line of thinking in biological systematics that deals with these groups, that is, Phylogenetic Systematics.

Cladistic analysis an analysis with specific methods in search for apomorphies, which in turn support monophyletic groups also known as clades. The outgroup method is the more frequently used in a cladistics analysis. The same as phylogenetic analysis.

Cladogram a dendrogram, in which the arguments for the organization of the branches are apomorphies.

Classes (of pre-evolutionary classifications) universal entities not delimited spatio-temporally, recognized according to their definition.

Coelacanths living subgroup of sarcopterygians, whose anterior dorsal fin is not fleshy. *Latimeria* is the living genus, that has lungs filled with fat and internal fecundation without a known intromittent organ, that preys mainly on cephalopods and moves their paired locomotor appendages alternately, and that lives on the continental platform in some regions around the Indian Ocean. Coelacanthiformes are also called Actinistia.

Convergence see adaptive convergence.

Cranium from Greek word *kraníon*, meaning cranium and head. From Latin, *cranium* (Houaiss and Villar 2009). Set of bones forming the head of a vertebrate. See also skull.

Crown group expression used in paleontology to define the smallest group that includes the living studied forms, and also their closest fossil relatives, all composing a monophyletic group.

DELTRAN Acronym for delayed transformation series. See further explanation under ACCTRAN.

Dendrogram any branched diagram.

Derived taxa in a phylogeny, the terminals that split later. It is not to be implied that they are more advanced or superior in any sense. In any case, each of them has its own autapomorphies and, therefore, has its own evolutionary innovations. Derived taxa may be equal to the crown group, but only by chance, not by definition. See also basal taxon and stem groups.

Dipnoans the living lungfishes, a subgroup of sarcopterygians. They present a durophagous buccal apparatus for grinding invertebrates with hard shells or carapaces, for food. This unique buccal apparatus is an apomorphy of the group. These animals currently live in tropical freshwater. The African and South American species are able to estivate throughout the dry season.

Dipnomorphs dipnoans (lungfishes) and their sister group, the †porolepiforms. Histological details of their scales are said to be apomorphies of the group. The increase in the number of mesomeres (bony segments) of the pectoral fin skeleton to more than four is what turns this fin into a leaf-shaped structure, typical of the group. Some other apomorphies have been proposed.

Dorsal fin(s) see median fins.

Elpistostegids as for †panderichthyids.

Evolutionary convergence the same as adaptive convergence.

Evolutionary step in the sense used in this book, it is any change in a feature that has been fixed in a species (no matter if by natural selection or by genetic drift), so that it can be passed on to descendant species. Gains, losses and/or reversals are evolutionary steps equally taken into account in phylogenetic analyses, and in this book. For instance, gain and loss of hairs are both evolutionary steps, treated equally as evolutionary steps (in this case, the loss may be considered as a reversal). Therefore, an evolutionary step is an apomorphy.

Exaptation a feature that was developed in a taxon by a given natural selection pressure for one use (called function) and that later in the evolutionary history of that taxon (that is, in a subset of that taxon) could acquire a different use (called an effect). Also, a feature that emerged without a function, as a by-product of the selection of another feature in a given taxon, and that may become useful (having an effect) in a sublevel (subgroup) of that taxonomic group. Further, exaptation may be used as a verb – indicating the process of cooptation by a new use –, or as a noun – indicating that the feature now has a new use or effect.

Extant taxon a taxonomic group having living representatives.

Feature as for character.

Fishes see Pisces.

Functional neck this feature corresponds to a freely mobile cervical region of the vertebral column, located just behind the skull and anterior to the pelvic girdle. To be fully functional, it depends on the disconnection of the skull from the pelvic girdle. In turn, this depends on the loss of several skull bones and the withdrawal of the notochord from the skull.

Foot see limb.

Fossil usually mineralized remains of biological organisms, including bones, foot prints, tracks, impressions on sediments, etc.

Fossil taxon an extinct taxonomic group.

Generality level the same as universality level. The inclusiveness of a character among a collection of taxa which are under cladistic analysis.

Geologic time scale represents the time axis since the generation of planet Earth. It is divided into eons, eras, periods, epochs and ages.

Gnathostomes vertebrates with up-and-down movable lower jaw. Gnathostomes include †placoderms, chondrichthyans, †acanthodians, and bony fishes (which in turn include tetrapods and, therefore, human beings).

Homology belonging to the same series of transformation of a feature.

Homoplastic features features found to have a relation or condition of homoplasy among each other in a phylogenetic analysis.

Homoplasy similar or even identical apomorphies may emerge independently in more than one terminal or taxon in the analysis. This situation is termed homoplasy to make the reader aware that the feature or feature state is an apomorphy for each taxon, independently. Homoplasies do no unite a group as monophyletic, but they may support the monophyly of independent groups in the cladogram. As an example, endotermy emerged independently in birds and mammals.

Incertae sedis undetermined position of a taxon in a classification or phylogenetic tree.

Individuals (in living systems) spatiotemporally localized entities, having the status of evolutionary units, characterized by shared exclusive ancestry, inferred from evolutionary novelties.

Ingroup the collective of terminals taken as the object of a cladistic analysis.

Limb both anterior and posterior limbs are formed by three major segments. The proximal includes but a single bony element (respectively humerus anteriorly, and femur posteriorly). The intermediary segment is composed by two bones (radius and ulna in the forelimb, and tibia and fibula in the hind limb). The third element is the foot, which includes the wrist in the anterior limb and the ankle in the posterior limb, and more distally, the toes or digits.

Lissamphibians living amphibians, the monophyly of which is in dispute, and may include some fossil taxa.

Lower jaw the mandible, formed mainly by the dentary bone, which supports the lower teeth of the mouth.

Lungs sac-like or bladder-like anatomical structures, that have permanent connections with the throat for air inhalation and exhalation. They are internally coated with vascular, respiratory tissue.

Macroevolution as redefined in this book, a collective exaptation of several features, working together to originate a new life style, with or without the help of the adaptation of other features at the same time. At the limit, a single exaptation may generate a new life style and can be responsible for macroevolution. Therefore, macroevolution is the generation of a new life style by means of exapation(s). This novel definition does not foreclose any other parallel definition, or alternative points of view on macroevolution. The original definition of macroevolution, as evolution above the species level, is not clear enough, once evolution can only occur in existing, real species, and, therefore, at the species level. This definition can now be dropped from the evolutionary literature. A function-free, duplicated gene may assume a new function (effect) and may represent a molecular case of macroevolution.

Median fins these are the locomotor appendages embedded in the sagittal line or midline, that divides the body into the right and left parts. The dorsal fin(s) lie along the middorsal line. The anal fin lies along the midventral line, behind the anus. The dorsal and anal fins are usually responsible for directing the movement of the animal from right to left, and vice-versa. The caudal fin lies at the confluence of the middorsal and midventral lines and is usually responsible for the propulsion of the animal.

Metapodial the proximal element of a toe or digit that is part of the palm or sole of anterior or posterior feet, termed a metacarpal or metatarsal, respectively.

Microevolution usually taken as evolution at the species level. As there is no other way of evolution to occur except at the species level, this definition seems to be misleading. Usually, microevolution involves the idea of step-by-step evolution by natural selection, as proposed by Darwin. Later it included genetic drift, mutation and gene flow, which occur through vast periods of geologic time. *Micro-* includes the idea of small or short. Therefore, microevolution is understood to be the set of evolutionary events that affect populations or species and that consist in gene frequencies and/or chromosome changes in the interval of a few generations. Microevolution therefore is linked to the idea of natural selection generating adaptation.

Monophyletic group in phylogenetic systematics it is defined by the presence of apomorphies. It includes all descendants of one exclusive common ancestral species, including this ancestor (in theory, as this ancestral species is now supposedly extinct). The same as clade.

Neck see functional neck.

Neotetrapoda tetrapods where the notochord does not develop (insert) inside the skull, in adults, among other apomorphies. By this definition, on the basis of current knowledge, Neotetrapoda (the neotetrapods) comprise a synonym of the so-called the crown-group tetrapods, together with †baphetids.

Notochord a rod-like structure, long, steady, but flexible, extending from head to tail, through the back of a vertebrate (or vertebrate-related organism). Usually it is formed by gelatinous cells surrounded by a sheath of fibrous connective tissue. In adult vertebrates, it tends to be substituted by a series of vertebrae.

Osteichthyans see bony fishes.

Outgroup taxon or group of taxa not pertaining to the ingroup and used in the out-group method of character comparison.

Outgroup comparison the same as out-group method.

Outgroup method Comparison between the taxa in focus (called collectively as the ingroup), with outside representatives (the outgroups). In evolutionary terms, the differences found in the ingroup in comparison with the several appropriate outgroups can only be interpreted as evolutionary innovations (novelties) for the ingroup: the apomorphies.

Outgroup rule Given two characters (or character states) that are homologues and found within a single monophyletic group, the character that is also found in the sister group is the plesiomorphic character, whereas the character found only within the monophyletic group is the apomorphic character (Wiley 1979: 139). We would like to highlight that the use of a single out-group (even if it is the sister group) may generate some technical artifacts. For instance, any autapomorphic state of the sister group is automatically considered as a plesiomorphy and therefore a plesiomorphic state for the ingroup may be mistakenly taken as an apomorphy in this case.

Paired locomotor appendages in this book, this expression refers to the vertebrate body appendages that occur in pairs and are used to move or direct the movement of the animal. In the human being, there is a pair of arms (that include the hands) and a pair of legs (that include the feet). Considered as homologous to these pairs of human appendages are the anterior and posterior legs (the limbs) of salamanders and pectoral and pelvic fins of sharks and golden fish, for instance.

Panderichthyids the fossil taxon that includes †*Elpistostege* and †*Panderichthys*; more recently it also includes †*Tiktaalik*. As each of these genera seems to be more closely related to Tetrapoda than to each other, †Panderichthyidae is thought to be a paraphyletic group. Sometimes, panderichthyids are alternatively called as †elpistostegids.

Paraphyletic group in Phylogenetic Systematics, it is the taxonomic group that does not includes all of the descendants of an ancestral species. For instance, Pisces is a paraphyletic group because it was originally erected to exclude Tetrapoda, which are indeed part of Pisces, in a definition that takes into account the proposition of apomorphies.

Parsimony in Science, it is the choice for the simplest hypothesis. In the sense of Phylogenetic Systematics, it is the hypothesis of the smallest number of evolutionary steps. This definition does not imply that evolution occurred by the way of the shortest number of evolutionary steps, but simply represents an impersonal methodological way to choose among several competing hypotheses. As a by-product, congruence is improved among all studied apomorphies. As another by-product of this impersonal way of choosing, there is the diminishing of *ad hoc* hypotheses of monophyletic groups and of character distribution in the final resulting phylogeny. Gains, losses and/or reversals are all considered evolutionary steps, equally counting in a parsimony analysis.

Pectoral fins see paired locomotor appendages.

Pelvic fins see paired locomotor appendages.

Phylogenetic analysis usually considered the same as cladistics analysis.

Phylogenetic Systematics a line of thinking in biological systematics that focuses on searching for feature modifications through time, as a means of delimiting each taxon, instead of using similarities and differences among taxa. The novelties that emerge are called apomorphies, and these novelties are the arguments for proposing natural groups in terms of evolutionary thinking. These natural groups are finally called monophyletic groups or clades (hence, the alternative name Cladistics for the Phylogenetic Systematics).

Phylogenetic tree in the sense of this book, a cladogram. More freely, any dendrogram expressing ancestral/descendent relationships among studied taxa.

Phylogeny as for phylogenetic tree.

Pisces a taxon that includes animals popularly called fishes, such as sharks and their relatives, the golden fishes and their relatives, coelacanths, lungfishes and sometimes also lampreys and hagfishes are included. As several taxa were erected prior to the advent of Phylogenetic Systematics, many of them were not found to be monophyletic groups, such as Pisces, for instance. Therefore, in evolutionary terms, Pisces does not exist.

Placoderms extinct group of vertebrates that may be defined by the presence of two independent shields on the body: the head and trunk shields. Their monophyly has been challenged.

Plesiomorphy a character state (or simply a character) that has a wider distribution than the studied level. Therefore, a plesiomorphic state is an apomorphy for a larger taxon and, vice versa, an apomorphy for a taxon is a plesiomorphy for any of its subgroups. Plesiomorphy is a useless concept in phylogenetic systematics (Cladistics), because in this approach of biological systematics, only derived states or new characters are meaningful for indicating exclusive common descendance or monophyletic lineages. The term is useful only to inform the reader that a given character or character state is not an apomorphy at that referred hierarchical level of a phylogeny. For instance, hair is an apomorphy for Mammalia (amongst living vertebrates) but a plesiomorphy for any of its subgroups, such as cats or Felidae, or any other subgroups of mammals.

Polarization (of a transformation series) determination of the supposed direction of a character transformation series, from the plesiomorphic to progressively more derived states.

Polyphyletic group grouping of terminals which indeed have different evolutionary origins. The grouping of homeotherm (indeed, endotherm) vertebrates in a taxon that could be named the "Homethermia" or "Endothermia" would be of polyphyletic nature: birds are derived dinosaurs and mammals are derived synapsids, for instance.

Polytomy a condition in a dendrogram where three or more branches emerge from a single point. In terms of phylogenetic systematics, it represents insufficient resolution, that is, poor knowledge on apomorphies that could depict the sister group relationships among the three lineages involved.

Radial (element of fins) The pectoral and pelvic fin rays are connected to their respective girdle by skeletal elements called radials. Median fins may have proximal, medial and distal radials for each fin ray, and collectively are called a pterygiophore.

Reptilomorphs part of the crown group Tetrapoda, cladistically more closely related to amniotes than to living amphibians (lissamphibians).

Rhipidistians sarcopterygians with a partially septate heart, among other apomorphies. It includes living forms such as lungfishes and tetrapods, as well as many fossil taxa, such as porolepiforms, "osteolepiforms", "panderichthyids" and others, such as tetrapods, including human beings.

Sarcopterygians vertebrates with intrinsic muscles in their paired locomotor appendages. Except for the most basal genera, all have a single proximal skeletal element in these paired locomotor appendages, that humans call humerus in their arms and femur in their legs. Sarcopterygians include the coelacanths, lungfishes and tetrapods, and also human beings.

Sister group a monophyletic group that shares an exclusive common ancestral species with another monophyletic group. One is the sister group of the other. "Sister group" is a designation that can only be given to a dichotomic situation in the cladogram. Where a trichotomy or polytomy is found, the resolution is considered incomplete and the designation "sister group" is inappropriate.

Skull The head skeleton. Roughly the same as cranium, but some authors do not include the lower jaw in the skull (Romer and Parsons 1977: 216). From Greek, *kranion* (Romer and Parsons 1977: 575). See also cranium.

Spandrel by-products of the emergence of features by natural selection were compared by Gould and Lewontin (1979) to the architectonic "spandrels" of the church of San Marcos in Venice (Italy), which were not created by architects for bearing paintings, but that were painted after the conclusion of the construction process, because these spandrels are available spaces for paintings. Therefore, in this sense, "spandrels" are evolutionary by-products of the emergence of other features. Thus, it is expected that spandrels have no function at their evolutionary origin. If later in the taxonomic history (that is, in a less inclusive taxon), they become somehow used, it is said that they were exapted, and now they have effects. See by-products, also for gene duplications.

Stem group a paraphyletic taxon that includes the close relatives of a given crown group. For example, the stem tetrapods are extinct taxa, each of them progressively more closely related to the crown group. Therefore, the definition of stem group is dependent on the definition of the crown group.

Stem tetrapods see stem group, crown group and Neotetrapoda.

Structure in this book, this term is often utilized as a synonym of character.

Synapomorphy an apomorphy of more than one terminal in the analysis, uniting them into a monophyletic group.

Systematics the field of biology that deals with the classification of organisms, the description of species and of supraspecific taxonomic groups, and how characters may be used in this regard. Taxonomy is hard to distinguish from systematics, but usually it is circumscribed to the rules and practices of describing taxa, segregated subjectively.

Systematization organizing taxa into a natural system of evolutionary taxa, a more adequate view based on System Theory for Phylogenetic Systematics.

Tarsus the set of small bones that compose the ankle.

Taxa plural form of taxon, as used in English.

Taxon a group of biological organisms, delimited by attributes. In Phylogenetic Systematics, these attributes are apomorphies, and the taxon is said to be a monophyletic group. The plural form of taxon is taxa.

Taxonomy see systematics.

Teleosts derived group of actinopterygians, with a homocercal caudal fin (upper lobe of caudal fin developed similarly to lower lobe). The lung becomes a gas bladder in this group of vertebrates. Most known representatives are tunafish, salmon, trout, cod, moray, eel, sardine, herring, aquarium tetras, catfish, swordfish, barracuda, scorpionfish, pufferfish, goosefish, batfish, stargazer, sunfish, reaching about 30,000 living species, the largest group of vertebrates.

Terminal every and each taxonomic group selected for an analysis in Phylogenetic Systematics (also called Cladistics). Terminals may be species if the researcher is studying the phylogeny of a group of species or any other supraspecific taxa, such as classes, for instance; in this case, each class is a terminal. The set of terminals is usually called the ingroup.

Terrestriality in this book, it has the meaning of walking, eating and breathing atmospheric air on dry land, totally independent of water, the animal not needing to return to an aquatic environment for any of these vital functions.

Tetrapodomorphs animals with choanae that are considered homologous among each other (said to be "true" choanae), but analogous to the choanae of lungfishes. Human beings are tetrapodomorphs.

Tetrapods vertebrates that are part of Sarcopterygii and have feet, including human beings.

Topology phylogenetic relationships amongst terminals, as shown by a cladogram.

Trait as for character.

Transformation series (of a feature) sequential modifications of a feature in a monophyletic lineage (monophyletic group). In the evolutionary paradigm, one feature is modified by DNA mutation, fixed in the species by means of natural selection or genetic drift and transmitted to the descendent species. If this very same modified feature is subject to another similar modification, the two states (modifications of a feature) are said to be homologous, and therefore belong to the same transformation series. Further, the original state (previous to these modifications) also belongs to this series. Each modification is an apomorphy that delimits a less inclusive hierarchical level in the monophyletic lineage. For instance, the feet appeared in Tetrapoda, with eight digits; when terrestriality eventually occurred, only five digits were present; feet with eight digits delimit Tetrapoda, five digits delimit the less inclusive Neotetrapoda, and four digits (in the hands) delimits an even less inclusive group in Batrachomorpha. All belong to the same transformation series, and delimit progressively less inclusive groups in Sarcopterygii.

Tree (phylogenetic) in the sense of this book, a cladogram.

Trichotomy specific condition of a polytomy with three branches emerging from a single point in a dendrogram.

Typological classification classifications by similarities and differences typical of Linnean taxonomy.

Universality level a hierarchical level in a phylogeny. In cladograms, delimited by apomorphies. Conversely, it is the level of distribution of a character or apomorphy. The same as generality level.

Upper jaw the place of contact with the mandible (lower jaw) when the mouth is shut, and it is formed by the premaxillary anteriorly and the maxillary, bones that frequently present teeth.

Vertebral centrum the cylinder-like part of the vertebra, which usually develops around the notochord, replacing it at least partially. Above the vertebral centrum emerges the neural arch, to protect the neural chord, from where emerge the nerves following to several regions of the body.

Vertebrates animals with a vertebral column, composed by vertebrae. Each vertebra is formed by a vertebral centrum and a neural arch above, from which emerges a neural spine.

References

Agosta SJ, Brooks DR. 2020. *The major metaphors of evolution: Darwinism then and now*. Cham: Springer.

Alfaro MK and 7 others. 2009. Nine exceptional radiations plus high turnover explain species diversity in jawed vertebrates. *Proceedings of the National Academy of Sciences* 106: 13410–13414.

Allen GE. 2014. The history of evolutionary thought. In JB Losos (ed) *The Princeton guide to evolution*. Princeton: Princeton University Press, pp. 10–27.

Andrews PW, Gangestad SW, Matthews D. 2002. Adaptationism – how to carry out an exaptationist program. *Behavior and Brain Sciences* 25: 489–553.

Baerends GP. 1971. The ethological analysis of fish behavior. In WS Hoar, DJ Randall (eds) *Fish Physiology. Vol. VI. Environmental relations and behavior*. New York: Academic Press, pp. 279–370.

Barnett SA, Fuller S. 1998. Dialogue review: Science. *Interdisciplinary Science Review* 23: 331–339.

Baum DA, Larson A. 1991. Adaptation Reviewed: a phylogenetic methodology por studying character macroevolution. *Systematic Zoology* 40: 1–18.

Beaumont EH, Smithson TR. 1998. The cranial morphology and relationships of the aberrant Carboniferous amphibian *Spatichephalus mirus* Watson. *Zoological Journal of the Linnean Society* 122: 187–209.

Benton MJ. 2005. *Vertebrate palaeontology*. 3rd ed. Malden: Blackwell Publishing.

Benton MJ. 2014. Causes and consequences of extinctions. In JB Losos (ed) *The Princeton guide to evolution*. Princeton: Princeton University Press, pp. 579–585.

Benton MJ. 2015. *Vertebrate Palaeontology*. 4th ed. Chichester: Wiley Blackwell.

Bittencout-dos-Santos W, El-Hani CN. 2013. A abordagem do pluralismo de processos e da evo-devo em livros didáticos de biologia evolutiva e zoologia de vertebrados. *Revista Ensaio* 15: 199–216.

Blackburn DG. 2002. Use of phylogenetic analysis to distinguish adaptation from exaptation. *Behavior and Brain Science* 25: 507–508.

Bond CE. 1979. *Biology of fishes*. Philadelphia: Saunders College Publishing.

Brazeau M. 2009. The braincase and jaws of a Devonian "acanthodian" and modern gnathostome origins. *Nature* 457: 305–308.

Brazeau M, Friedman M. 2014. The characters of the Paleozoic jawed vertebrates. *Zoological Journal of the Linnean Society* 170: 779–821.

Brusca RC, Brusca GJ. 2007. *Invertebrados*. 2nd. ed. Rio de Janeiro: Editora Guanabara Koogan S. A.

Burrow C, den Blaauwen J, Newman M, Davidson R. 2016. The diplacanthid fishes (Acanthodii, Diplacanthiformes, Diplacanthidae) from the Middle Devonian of Scotland. *Palaeontologia Electronica* 19.1.10A: 1–83.

Burrow CJ, Turner S. 2010. Reassessment of "*Protodus*" *scoticus* from the Early Devonian of Scotland. In DK Elliott, JG Maysey, X Yu, D Miao (eds), *Morphology, phylogeny and paleobiogeography of fossil fishes*. Munich: Verlag Dr. Friedrich Pfeil, pp. 123–144.

Byrne HM, Green M, Ahlberg PE, Balbus SA. 2018. Tides: a key driver in the evolution of terrestrial vertebrates? PL41A-02. Portland: *Ocean Sciences Meeting of The Oceanographic Society*.

Carroll RL. 1988. *Vertebrate Paleontology and Evolution*. New York: W. H. Freeman and Company.

Carroll RL, Irwin J, Green DM. 2005. Thermal physiology and the origin of terrestriality in vertebrates. *Zoological Journal of the Linnean Society of London* 143: 345–358.

Casinos A. 2017. From Cuénot's *préadaptation* to Gould and Vrba's *exaptation*: a review. *Biological Journal of the Linnean Society* 121: 239–247.

Charlton BG. 2008. Zombie Science: a sinister consequence of evaluating scientific theories purely on the basis of enlightened self-interest. *Medical Hypotheses* 71: 327–329.

Christoffersen ML. 2001. Ciência ultramoderna. *Conceitos* 4: 64–69.

Christoffersen ML. 2011. Evolução (Darwin) e Sistemática Filogenética (Hennig): Paradigmas comensuráveis. *Conceitos* 16: 113–120.

Clack JA. 2002. *Gaining ground: The origin and evolution of tetrapods*. Bloomington: Indiana University.

Clack JA. 2012. *Gaining ground: The origin and evolution of tetrapods*. 2nd ed. Bloomington: Indiana University Press.

Clack JA and 14 others. 2016. Phylogenetic and environmental context of Tournaisian tetrapod fauna. *Nature Ecology and Evolution* 1: 1–11.

Clack JA, Ahlberg PE. 2016. Sarcopterygians: From lobe-finned fishes to the tetrapod stem group. In JA Clack, R Fay, A Popper (eds) *Evolution of the Vertebrate Ear – Evidence from the Fossil Record, Springer Handbook of Auditory Research vol. 59*. Springer, Cham, pp. 51–70.

Clouthier R, Ahlberg PE. 1996. Morphology, characters, and the interrelationships of basal sarcopterygians. In MLJ Stiassny, LR Parenti, GD Johnson (eds) *Interrelationships of fishes*, New York: Academic Press, pp. 445–479.

Coddington JA. 1988. Cladistic tests of adaptational hypotheses. *Cladistics* 4: 3–22.

Cohen MR. (ed) 1923. *Change, logic, love: philosophical essays by the late Charles Sanders Peirce*. London: Kegan Paul.

Cuénot L. 1901. L'évolution des théories transformistes. *Revue générale des Sciences pures et appliquées* 12: 264–269.

Cuénot L. 1909. Le peuplement des places vides dans la nature et l'origine des adaptations. *Revue générale des Sciences pures et appliquées* 20: 8–14.

Cuénot L. 1914. Théorie de la préadaptation. *Scientia* 16: 60–73.

Currie PJ. 1981. *Hovasaurus boulei*, an aquatic eosuchian from the upper Permian of Madagascar. *Palaeontologica Africana* 24: 99–168.

Daeschler EB, Shubin NH, Jenkins Jr FA. 2006. A Devonian tetrapod-like fish and the evolution of the tetrapod body plan. Nature 440: 757–763.

Darwin CR, Wallace A. 1858. On tendencies of species to form varieties; and on the perpetuation of varieties and species by natural means of selection. *Journal of the Proceedings of the Linnean Society* 3: 45–62.

Darwin CR. 1859. *On the origin of species by means of natural selection, or the preservation of favored races in the struggle for life*. London: Murray.

De Carvalho MR. 1996. Higher-level elasmobranch phylogeny, basal squaleans, and paraphyly. In MLJ Stiassny, LR Parenti, GD Johnson (eds) *Interrelationships of fishes*, New York: Academic Press.

De Pinna MCC. 1991. Concepts and tests of homology in the cladistic paradigm. *Cladistics* 7: 367–394.

De Pinna MCC. 1998. Phylogenetic relationships of Neotropical Siluriformes: Historical overviews and synthesis of hypothesis. In LR Malabarba, RE Reis, RP Vari, CAS Lucena, ZMS Lucena (eds) *Phylogeny and classification of Neotropical fishes*. Porto Alegre: EDIPUCRS, pp. 279–330.

Dean B. 1895. *Fishes Living and Fossil*. London: MacMillan and Co.

Dickson B, Clack JA, Smithson TR, Pierce SE. 2020. Functional adaptive landscapes predict terrestrial capacity at the origin of limbs. *Nature* https://doi.org/10.1038/s41586-020-2974-5.

Dickson B, Clack JA, Smithson TR, Pierce SE. 2021. Functional adaptive landscapes predict terrestrial capacity at the origin of limbs. *Nature* 589: 242–245.

Dupret V, Sanchez S, Goujet D, Tafforeau P, Ahlberg PE. 2014. A primitive placoderm sheds light on the origin of the jawed vertebrate face. *Nature* 507: 500–503.

Eigenmann CH. 1921. The American Characidae. *Memoirs of the Museum of Comparative Zoology* 43, pt. 3: 209-310 + pls 30-92.

Ellis EA and 9 others. 2022. Sexual signals persist over deep time: ancient co-option of bioluminescence for courtship displays in Cypridinid Ostracods. *Systematic Biology* 2022: 1–11.

Ferreira ABH. 1999. *Novo Aurélio. O dicionário da língua portuguesa século XXI*. Rio de Janeiro: Editora Nova Fronteira.

Fraas E. 1889. Die Labyrinthodonten der Schwäbischen Trias. *Palaeontographica* 36: 1–158.

Freeman S, Herron JC. 2001. *Evolutionary Analysis*. 2nd ed· Upper Saddle River: Prentice Hall.

Frenkel-Pinter M, Petrov AS, Matange K, Travisano M, Glass J, Williams LD. 2022. Adaptation and exaptation: from small molecules to feathers. *Journal of Molecular Evolution* 90: 166–175.

Fricke H, Reinicke O, Hofer H, Nachtigall W. 1987. Locomotion of the coelacanth *Latimeria chalumnae* in its natural environment. *Nature* 329: 331–333.

Friedman M. 2015. The early evolution of ray-finned fishes. *Palaeontology* 58: 213–228.

Friedman M, Giles S. 2016. Actinopterygians: the ray-finned fishes – an explosion of diversity. In JA Clack, R Fay, A Popper (eds) *Evolution of the Vertebrate Ear – Evidence from the Fossil Record*, Springer Handbook of Auditory Research vol. 59. Springer, Cham, pp. 17–49.

Fuller S. 1998. *Science*. London: Open University Press.

Fuller S. 1999. Is Science studies lost in the Kuhnian plot? On the way back from paradigms to movements. *Science as Culture*, 8(4): 405–435.

Futuyma DJ. 1995. *Biologia Evolutiva*. 2ª. ed. Ribeirão Preto: Sociedade Brasileira de Genética.

Gaffney ES. 1979. Tetrapod Monophyly: a phylogenetic analysis. *Bulletin of the Carnegie Museum of Natural History*, 13: 92–105.

Géry J. 1977. *Characoids of the World*. Neptune City: T. F. H. Publications.

Goodrich ES. 1909. *Vertebrata Craniata, Treatise on Zoology*, R Lankester (ed), Pt. 9 London: Adam and Charles Black.

Goodrich ES. 1930. *Studies on the Structure and Development of Vertebrates*. Chicago: University of Chicago Press.

Goujet D. 2011. "Lungs" in Placoderms, a persistent palaeobiological myth related to environmental preconceived interpretations. *Comptes Rendus de Palevolution* 10: 323–329.

Goujet D, Young GC. 2004. Placoderm anatomy and phylogeny: new insights. In G Arratia, MVH Wilson, R Cloutier (eds) *Recent advances in the orign and early radiation of vertebrates*. München: Verlag Dr. Friedrich Pfeil, pp 109–126.

Gould SJ. 1984. Only his wings remained. *Natural History* 93: 10–18.

Gould SJ. 1997. Evolutionary psychology: An exchange. *New York Review of Books* 44: 56–58.

Gould SJ, Lewontin RC. 1979. The spandrels of San Marco and the Panglossian paradigm: a critique of the adaptationist programme. *Proceedings of the Royal Society of London* 205: 581–598.

Gould SJ, Vrba SE. 1982. Exaptation – a missing term in the Science of form. *Paleobiology* 8: 4–10.

Gregory WK. 1933. Fish Skulls: a study of the evolution of natural mechanisms. *Transactions of the American Phylosophical Society* 23: 75–481.

Gutarra S, Rahman IA. 2022. The locomotion of extinct secondarily aquatic tetrapods. *Biological Reviews* 97: 67–98.

Harder W. 1975. *Anatomy of fishes. Part II figures and plates*. Stuttgart: E. Schweizerbart'sche Verlagsbuchhandlung.

Harmon LJ. 2014. Macroevolutionary rates. In JB Losos (ed) *The Princeton guide to evolution*. Princeton: Princeton University Press, pp. 567–572.

Harris VA. 1960. On the locomotion of the mud-skipper *Periophthalmus koelreuteri* (Pallas): (Gobiidae). *Proceedings of the Zoological Society of London* 134: 107–135.

Henderson S, Dunne EM, Fasey SA, Giles S. 2022. The early diversification of ray-finned fishes (Actinopterygii): hypotheses, challenges and future prospects. *Biological Reviews*, 2022: 284–315 (https://doi.org/10.1111/brv.12907).

Hennig W. 1950. *Grundzüge einer Theorie der phylogenetischen Systematik*. Berlin: Deutscher Zentralverlag.

Hennig W. 1966. *Phylogenetic systematics*. Urbana: University of Illinois Press.

Hennig W. 1975. "Cladistic analysis or cladistics classification?" A reply to Mayr. *Systematic Zoology* 24: 244–256.

Hildebrand M. 1982. *Analysis of Vertebrate Structure*. 2nd ed. New York: Wiley.

Houaiss A, Villar M de S. 2009. *Dicionário Houaiss da língua portuguesa*. 1ª. Ed. Rio de Janeiro: Objetiva.

Human BA, Owen EP, Compagno LJV, Harley EH. 2006. Testing morphologically based phylogenetic theories within the cartilaginous fishes with molecular data, with special reference to the catshark family (Chondrichthyes: Scyliorhinidae) and the interrelationships within them. *Molecular phylogenetics and evolution* 39: 384–391.

Jablonski D. 2005. Mass extinctions and macroevolution. *Paleobiology* 31: 192–210.

Janvier P. 1996. *Early vertebrates*. Oxford: Clarendon Press.

Janvier P. 2004. Wandering nostrils. *Nature* 432: 23–24.

Johansen K. 1970. Air breathing in fishes. In WS Hoar, DJ Randall (eds) *Fish Physiology. Vol. IV. The nervous system, circulation, and respiration*. San Diego: Academic Press, pp 361–411.

Jordan DS, Evermann BW. 1908. *American food and game fishes. A popular account of all the species found in America north of the Equator, with keys for ready identification, life histories and methods of capture*. New York: Doubleday, Page and Company.

Keller EF. 1993. The paradox of scientific subjectivity. *Annals of Scholarship* 9: 135–153.

King B, Quiao T, Lee MSY, Zhu M, Long JA. 2016. Bayesian Morphological Clock Methods Ressurrect Placoderm Monophyly and Reveal Rapid Early Evolution in Jawed Vertebrates. *Systematic Biology* 66: 499–516.

Kingsley JS. 1899. *Text Book of Vertebrate Zoology*. New York: Henry Holt and Company.

Kivell TL, Begun DR. 2007. Frequency and timing of scaphoid-centrale fusion in hominoids. *Journal of Human Evolution* 52: 321–340.

Kong H. and 5 others 2007. Patterns of gene duplication in the plant *SKP1* gene family in angiosperms: evidence for multiple mechanisms of rapid gene birth. *The Plant Journal* 50: 873–885.

La Porta CAM, Zapperi S, Pilotti L. 2020. *Undestanding Innovation through exaptation*. Cham: Springer.

Lagler KF, Bardach JE, Miller RR. 1962. *Ichthyology*. New York: Wiley.

Lankester ER, Ridewood WG. 1908. *Guide to the gallery of fishes in the department of zoology of the British Museum (Natural History)*. London: Printed by order of the trustees of the British Museum.

Larson A. 2014. Concepts in character macroevolution: adaptation, homology, and evolvability. In JB Losos (ed) *The Princeton guide to evolution*. Princeton: Princeton University Press, pp. 89–99.

Lauder GV, Liem KF. 1983. The evolution and interrelationships of the actinopterygian fishes. *Bulletin of the Museum of Comparative Zoology Harvard University* 150: 95–197.

Laurin M. 2010. *How Vertebrates Left the Water*. Berkeley: University of California Press.

Lewontin R. 1978. Adaptation. *Scientific American* 239: 212–230.

Lewontin R. 1979. Sociobiology as an adaptationist program. *Behavioral Science* 24: 5–14.

Linde-Medina M. 2020. On the problem of biological form. *Theory in Biosciences* 139: 299–308.

Lindsey CC. 1978. Form, function, and locomotory habits in fish. In Hoar WS, DJ Randall (eds) *Fish Physiology, Vol. VII. Locomotion*. New York: Academic Press, pp 1–100.

Long JA. 2011. *The rise of fishes: 500 million years of evolution*. 2nd ed. Baltimore: The Johns Hopkins University Press.

Long JA. and 11 others 2015. Copulation in antiarch placoderms and the origin of gnathostome internal fertilization. *Nature* 517: 196–199.

Losos JB. 2014. What is evolution? In JB Losos (ed) *The Princeton guide to evolution*. Princeton: Princeton University Press, pp. 3–9.

Luksevics E, Zupins I. 2004. Sedimentology, fauna and taphonomy of the Pavari site, Late Devoninan of Latvia. *Acta Universitatis Latviensis* 679: 99–119.

Maddison WP, Donoghue MJ, Maddison DR. 1984. Outgroup analysis and parsimony. *Systematic Zoology* 33: 83–103.

Maderson, PFA. 1972. When? Why? And How?: some speculations on the evolution of the vertebrate integument. *American Zoologist* 12: 159–172.

Maisch MW. 2010. Phylogeny, systematics, and origin of Ichthyosauria – the state of the art. *Palaeodiversity* 3: 151–214.

Maldanis L. and 18 others. 2016. Heart fossilization is possible and informs the evolution of cardiac outflow tract in vertebrates. *eLife*, 5: 1–12.

Mayr E. 1974. Cladistic analysis or cladistic classification? In E Mayr (ed) *Evolution and the diversity of life. Selected essays*. Cambridge: The Belknap Press, pp. 431–475.

McFarland WN, Pough FH, Cade TJ, Heiser JB. 1979. *Vertebrate Life*. 1st ed. New York: Macmillan Publishing Company.

McGhee GR. 2013. *When the invasion of land failed: The legacy of the Devonian extinctions*. New York: Columbia University Press.

McLennan DA. 2008. The concept of co-option: why evolution often looks miraculous. *Evolution: Education & Outreach* 2008: 247–258.

McNamara K, Selden P. 1993. Strangers on the shore. *New Scientist* 139: 23–27.

Merton RK. 1938. *Science, technology and society in seventeenth century England.* New York: Howard Fertig.

Moy-Thomas JA, Miles RS. 1971. *Palaeozoic Fishes*. London: Chapman and Hall, Ltd.

Moyle PB, Cech JJ. 1996. *Fishes: an introduction to ichthyology*. 3rd ed. Upper Saddle River: Prentice Hall.

Nelson DO, Heath JE, Prosser CL. 1984. Evolution of temperature regulatory mechanisms. *American Zoologist* 24: 791–807.

Nelson JS. 1976. *Fishes of the world.* New York: John Wiley and Sons.

Nelson JS. 1984. *Fishes of the world.* 2nd ed. New York: John Wiley and Sons.

Nelson JS. 1994. *Fishes of the world.* 3rd ed. New York: Wiley.

Nelson JS. 2006. *Fishes of the world.* 4th ed. Hoboken: John Wiley and Sons.

Nelson JS, Grande TC, Wilson MVH. 2016. *Fishes of the World.* 5th ed. Hoboken: John Willey and Sons.

Newman HH. 1923. *Vertebrate Zoology*. New York: The Macmillan Company.

Niedzwiedski G, Szrek P, Narkiewicz K, Narkiewick M, Ahlberg PE. 2010. Tetrapod trackways from the early Middle Devonian period of Poland. *Nature* 463: 43–48.

Nilsson DE, Pelger S. 1994. A pessimistic estimate of the time required for an eye to evolve. *Proceedings of the Royal Society of London B* 256: 53–58.

Oyama S. 1991. The conceptualization of nature. Nature as design. In NJ Thompson (ed) *Gaia 2: Emergence: The new Science of becoming*. Hudson, NY: Lindsfarme Press, pp. 171–184.

Oyama S. 1993. Constraints and development. *Netherlands Journal of Zoology* 43: 6–16.

Oyama S. 1994. Rethinking development. In PK Bock (ed) *Handbook of psychological anthropology*. Westport, London: Greenwood Press, pp. 185–196.

Oyama S. 1999. Locating development: Locating developmental systems. In EK Scholnick, K Nelson, SA Gelman, PH Miller (eds) *Conceptual development. Piaget's legacy*. Hillsdale, NJ: Lawrence Erlbaum, pp. 185–208.

Oyama S. 2001. What do you do when all the good words are taken? In S Oyama, PE Griffiths, RD Gray (eds) *Cycles of contingency: Developmental systems and evolution*. Cambridge, MA: Massachusetts Institute of Technology Press, pp. 177–193.

Panchen AL. 1991. The early tetrapods: classification and the shapes of cladograms. In H-P Schultze, L Trueb (eds) *Origins of the higher groups of tetrapods: controversy and consensus*. Ithaca: Comstock Publishing Associates, pp. 110–144.

Papavero N. 1994. *Fundamentos Práticos de Taxonomia Zoológica*, 2a. ed. São Paulo: Editora Unesp.

Pardo JD, Szostakiwskyj M, Alhberg PE, Anderson JS. 2017. Hidden morophological diversity among early tetrapods. *Nature* 546: 642–646.

Pievani T, Serrelli E. 2011. Exaptation in human evolution: how to test adaptive *vs* exaptive evolutionary hypotheses. *Journal of Anthropological Sciences* 89: 9–23.

Pough FH, Janis CM. 2019. *Vertebrate life*. 10th ed. New York: Oxford University Press.

Pough FH, Janis CM, Heiser JB. 2005. *Vertebrate life*. 7th ed. Upper Saddle River: Pearson Prantice Hall.

Qiao T, King B, Long JA, Ahlberg PE, Zhu M. 2016. Early Gnathostome Phylogeny Revisited:

Multiple Method Consensus. *PLoS ONE* 11: e0163157.

Rade C. 2013. Functional morphology of aquatic substrate-based locomotion in walking batfishes (Lophiiformes; Ogcocephalidae). ProQuest Dissertations and Theses, Ann Arbor, http://search.proquest.com/docview/1372277467?accountid=134127.

Reeve HK, Sherman PW. 1993. Adaptation and the goals of evolutionary research. The *Quarterly Review of Biology* 68: 1–32.

Ridley M. 2004. *Evolution*. 3rd. Ed. Malden: Blackwell Publishing Company.

Romer AS. 1966. *Vertebrate paleontology*. 3rd ed. Chicago: University of Chicago Press.

Romer AS, Parsons TS. 1977. *The Vertebrate Body*. 5th ed. Philadelphia: W. B. Saunders Company.

Roth C. and 6 others 2007. Evolution after gene duplication: mechanisms, sequences, systems and organisms. *Journal of Experimental Zoology (Molecular and Developmental Evolution)* 308B: 58–73.

Sanchez S, Tafforeau P, Clack JA, Ahlberg PE. 2016. Life history of the stem tetrapod *Acanthostega* revealed by synchrotron microtomography. *Nature* 537: 408–411.

Sayer MDJ, Davenport J. 1991. Amphibious fish: why do they leave water? *Reviews in Fish Biology and Fisheries* 1: 159–181.

Schoch RR. 2014. *Amphibian evolution. The life of early land vertebrates*. Oxford: Wiley Blackwell.

Shelley C. 1999. Preadaptation and the explanation of human evolution. *Biology and Philosophy* 14: 65–82.

Shimizu D, Macho GA. 2007. Functional significance of the microstructural detail of the primate dentino-enamel junction: A possible example of exaptation. *Journal of Human Evolution* 52: 103–111.

Shirai S. 1996. Phylogenetic interrelationships of neoselachians (Chondrichthyes: Euselachii). In MLJ Stiassny, LR Parenti, GD Johnson (eds) *Interrelationships of fishes*, New York: Academic Press, pp 9–34.

Shubin NH, Daeschler EB, Jenkins Jr. F. 2014. Pelvic girdle and fin of *Tiktaalik roseae*. *Proceedings of the National Academy of Sciences of the United States of America* 111: 893–899.

Simpson GG. 1953. *The Major Features of Evolution*. New York: Simon and Schuster.

Standen EM, Du TY, Larsson HC. 2014. Developmental plasticity and the origin of tetrapods. *Nature* 513: 54–58.

Straile WE. 1960. Sensory hair follicles in mammalian skin: the tylotrich follicle. *American Journal of Anatomy* 160(2): 133–147.

Straile WE. 1969. Vertical cutaneous organization. *Journal of Theoretical Biology*, 24(2):203–215.

Tattersall I. 2004. What happened in the origin of human consciousness? *The Anatomical Record B* 276B: 19–26.

Thewissen JGM. 2014. *The walking whales: from land to water in eight million years.* Oakland: University of California Press.

Triques ML, Christoffersen ML. 2009. Exaptations in the conquest of land by Tetrapoda. *Gaia Scientia* 3: 69–74.

Triques ML, Chirstoffersen ML. 2015. Use of an explicit method for distinguishing exaptations from adaptations at the origin of terrestriality in Tetrapoda, based on cladograms and synapomorphies. *Gaia Scientia* 9: 179–189.

Triques ML, Chirstoffersen ML. 2017. *Peixes conquistam a terra firme: nova abordagem para um evento acidental único.* São Paulo: Editora Baraúna.

Triques ML, Chirstoffersen ML. 2018. Arguments for replacing the concept of *preadaptation* by *exaptation* at the origin of terrestriality in Vertebrata. *Biological Journal of the Linnean Society* 123: 235–246.

Uken R, Green AN. 2006. Coelacanth (*Latimeria chalumnae*) resting traces in a cave floor, Chaka Canyon, South Africa. *South African Journal of Science* 102: 474–475.

Voordeckers K. and 6 others 2012. Reconstruction of ancestral methabolic enzymes reveal molecular mechanisms underlying evolutionary innovation through gene duplication. *PLOS Biology* 10: 1–17.

Vorobyeva EI. 2008. The operational approach to the problem of tetrapodization. *Paleontological Journal* 42: 114–126.

Weitzman SH. 1962. The osteology of *Brycon meeki*, a generalized characid fish, with an osteological definition of the family. *Stanford Ichthyological Bulletin* 8: 1–77.

Wiens JJ, Brandley MC, Reeder TW. 2006. Why does a trait evolve multiple times within a clade? Repeated evolution of snakelike body form in squamate reptiles. *Evolution* 60: 123–141.

Wiley EO. 1981. *Phylogenetics. The Theory and Practice of Phylogenetic Systematics.* New York, John Wiley and Sons.

Wiley EO, Siegel-Causey D, Brooks DR, Funk VA. 1991. *The Compleat Cladist. A Primer of Phylogenetic Procedures.* Lawrence: The University of Kansas.

Williams GC. 1966. *Adaptation and natural selection.* Princeton: Princeton University Press.

Woodward AS. 1898. *Outlines of Vertebrate Palaeontology for Students of Zoology.* Cambridge: Cambridge University Press.

Zhu M, Yu XB. 2009. Stem sarcopterygians have primitive polybasal fin articulation. *Biology Letters* 5: 372–375.

Zhu M, Yu X, Ahlberg PE, Choo B, Lu J, Qiao T, Qu Q, Zhao W, Jia L, Blom H, Zhu, Y. 2013. A Silurian placoderm with osteichthyan-like marginal jaw bones. *Nature* 502: 188–194.

Zittel K von. 1911. *Grundzüge der Paläontologia* (*Paläozoologie*). *II Abteilung* VERTEBRATA. München und Berlin: Druck und Verlag Von R. Oldenbourgh.

Index

a single bone connecting paired locomotor appendages to girdles 84
absence of correlation among characters 148
acanthodians 61, 64–66, 98, 127, 159, 201
Acanthodii 5, 6, 49, 64–66, 138, 208
Acanthostega 50, 86, 88, 94, 108, 117, 122, 146, 165, 215
ACCTRAN 119, 197, 199
actinopterygians 7, 46, 53–58, 65, 73, 86, 91–94, 96, 124, 127, 143, 147, 154, 155, 182, 185, 198, 206
Actinopterygii 5, 7, 9, 16, 18, 25, 27, 36, 46, 48, 49, 52, 54, 56–59, 68, 73, 77, 86, 103, 110, 127, 128, 138, 147, 154, 182, 212
ad hoc
  explanations 28, 31, 161, 183, 204
adaptationism 33, 34, 179, 208
adaptationists XI, 79, 80, 163, 164, 178
adaptive
  should be favored immediately 142
adaptive scenarios 20, 44, 52, 58, 71, 141, 151, 156, 157, 163
addition on the functions 167
Aïstopoda 117, 118, 134
Amia 54, 91–93
Amniota 9, 46, 77, 117, 137
analogical reasoning 194
Anthracosauria 77
apomorphy 25, 68, 93, 116, 119, 139, 141, 146, 149, 180, 198–201, 203, 204, 206, 207
Arapaima gigas 26, 54
Archeria 118
arthropods 24, 30, 157
asynchrony
  as argument 148
Aytonerpeton 120

Baphetidae 72, 108, 111, 113, 118, 121, 147, 153, 158, 182–184
baphetids 121
basal 15, 26, 27, 46, 54–58, 65, 71, 72, 83, 85, 88, 94, 100, 102, 103, 108, 110, 117, 118, 120, 121, 125, 147, 155, 173, 182, 183, 185, 197, 200, 205, 209, 210
Batrachomorpha 71, 207
Blackburn
  method to distinguish exaptation from
    adaptation XIV, 34, 68, 69, 72, 79, 86, 89, 90, 94, 96, 97, 100, 102, 103, 105–108, 111, 113, 114, 120, 121, 123, 125, 126, 130–133, 144, 148, 149, 152–154, 156–160, 173–175, 186, 208
Bothriolepis 62, 63, 90, 140
burden of proof 44, 132, 162
by-product 37, 95, 96, 130, 170, 172, 198, 200, 204

Carboniferous 21, 24–26, 30, 31, 50, 52, 53, 55, 58, 59, 65, 94, 100, 102, 104, 117, 122, 134, 137–139, 142, 144, 154, 208
Cetacea 47
chance event of emergence of each and all necessary features 42
Characiformes 7, 126
Cheirolepis 57, 58
Chelonia 9, 47
choanae 73, 94, 105, 122, 143, 144, 151, 156, 168, 183, 184, 207
chondrichthyans 5, 15, 46, 48, 66, 67, 98, 127, 147, 201
Chondrichthyes 5–7, 49, 65, 66, 128, 156, 212, 215
Cladistics 1, 68, 198, 199, 204, 206, 210
Coelacanthiformes 9, 71–73, 77, 86, 113, 199
coelacanths 3, 9, 18, 26, 94, 183, 204, 205
collectively X, 42, 147, 184, 203, 205
Colosteidae 108, 117, 118
column 83
column rigidity 89
constructivist-interactionist-systemic 194
co-optation 36
coopted 35, 36, 81, 86, 116, 123, 127, 129, 141, 168, 175, 179
coopted effect 81
cooption
  regarding molecules 166
Copella arnoldi 28
cranium 11, 12, 41, 143, 144, 149, 157, 199, 205
  skull 41

Crassigyrinidae 102, 108, 117, 118
Crassigyrinus 72, 102
criteria
in Andrews et al. 2002 68
crown group 30, 53, 55, 57–59, 71, 72, 77, 108, 110, 111, 113, 117, 118, 120, 121, 140, 147, 152–156, 158, 161, 182, 183, 186, 198, 200, 205, 206
cyclical Science 194

Darwin IX, 1, 2, 16, 33, 34, 44, 132, 135, 136, 144, 145, 151, 153, 163, 164, 176, 178, 181, 192, 202, 209, 210
Darwinian theory is inadequate to explain major features of evolution 39
DELTRAN 119, 197, 199
Densignathus 103
depressiform fish 48
depressiform head and body 100
Devonian X, 21, 23, 25–27, 30, 49, 50, 52, 53, 55, 56, 58, 59, 64, 65, 103, 104, 107, 128, 139–141, 145–149, 155, 156, 184, 193, 197, 208–210, 213, 214
Diploradus 120
Dipnoi 9, 28, 51, 54, 71, 73, 86, 94, 96, 113, 138, 141, 183
Dipnomorpha 71
diversification 39, 52, 54–56, 58, 59, 103, 137, 154, 212
droughts 28
dynamical processes 195

effect X, 35–38, 42, 44, 72, 80, 81, 98, 102, 108, 116, 119, 123, 125, 128–130, 132, 135, 137, 140, 157, 165–167, 198, 200, 202
Elginerpeton 71
Elpistostege 50, 103, 203
elpistostegids 27
Embolomeri 118, 120
endochondral 18
Eryops 89
eurypterid 30
Eurypteridae 29
Eusthenopteron 46, 55, 72, 86, 125, 127
evolution proceeds at the species level 40
evolutionary advantages 24, 45, 99, 143
evolutionary pressures 28, 29, 45, 53, 116
evolutionary process by adaptation must be brief 142
evolutionary processes automatically represent \ 40
exaptation
does not endorse creationism 165
exaptation is not preadaptation 163
exaptation program 79
exaptation, an abrupt change in use 167
extrascapular 12, 19, 168

Famennian 53
feet 9, 16, 27, 45, 49, 50, 51, 80, 82, 86, 103, 105, 106, 114, 116, 122, 124, 125, 130, 131, 134, 137, 142–145, 147, 149, 150, 153–155, 171, 173, 184, 202, 203, 207
femur 83
first animal interpreted as being terrestrial 122
foot 9, 44, 49, 50, 79, 83, 86, 103, 106, 114, 116, 127, 129, 131, 138, 154, 166, 168, 171, 173, 201
Frasnian 25, 30, 52, 53, 55, 154
function 35

geologic time gap 175
Givetian 25, 30, 53
Gnathostomata 5, 11, 64, 140, 158

hagfishes 5
Hennig XIV, XV, 1, 2, 38, 160, 162, 192, 196, 209, 212
Hovasaurus 47
humerus 83
hydrodynamic vertebrates 45
Hynerpeton 103
hypotheses of exaptations and macroevolution
are testable 158

Ichthyosauria 47
Ichthyostega 50, 77, 89, 108, 117, 197
Ichthyostegalia 103
ichthyostegids 131
incertae sedis 7, 67
integrative Science 195
interopercle 12
intrinsic muscles 83
inverted cladogram 108
Ivorian 138

Koilops 120

lampreys 5
Latimeria 73, 83, 96, 199, 210, 216
Latin expressions XIX
Lepidosiren 28, 73, 83, 92, 96
lepidotrichiae 18
Lepisosteus 54, 92
Lepospondyli 71, 77, 117
Lethiscus 117, 134
life in shallow waters 100, 184
lifestyle 42, 107, 118, 119, 120, 124, 131, 182, 184
limb-reduced \"ecomorphs\" 49
Lissamphibia 3, 9, 77, 106, 197
Lophiidae 48
loss of opercular bones 99
loss of the series of opercular bones and of the extrascapular 143
lungfish 9, 26, 51, 73
lungs 25, 26, 28, 34, 54–56, 62, 73, 90–92, 94, 97, 98, 100, 103, 104, 107, 116, 126, 139, 140–145, 147, 150, 151, 155, 156, 165, 166, 171, 175, 182, 184, 187, 199

macroevolution X, 21, 38, 39, 40, 41, 42, 49, 51, 68, 71, 73, 115, 122, 123, 127, 129, 133, 135, 138–140, 144, 145, 148, 149, 151–153, 155, 156, 158–160, 162, 163, 165, 168, 185, 198, 202, 208, 212, 213
  a new alternative possibility of interaction with the environment 40
  an abrupt change of use of one or more features 38
  as originally proposed, loses its meaning 40
  correlated to a series of rapid transformations 39
  gene duplication event 38
  more noticeable when there is a collective exaptation 41
  new life style or new mode of life 41
  pluralistic interactions 39
  rupture 40, 42
macroevolution is exaptation 41
macroevolutionary processes IX, XII, 42
McLennan X, 68, 69, 214
Mesosauridae 48
Metaxygnathus 71, 103
microevolution X, 21, 24, 34, 36, 38, 39, 40, 42, 45, 46, 50, 68, 104, 114–116, 122, 135, 138–140, 143, 145, 149–152, 155, 158, 159, 162, 168, 202
misunderstandings on exaptation 123
monophyletic XIX, 1, 5–7, 17, 30, 40, 44, 45, 52, 57, 60, 62, 63, 65, 77, 90, 128, 133, 152, 159, 162, 173, 180, 183, 197, 198, 199, 201, 203–207
monophyletic groups
  represent real entities, individual systemas 40
Mosasauria 47
Moythomasia 56, 57
mudskipper 8
Myxinoidea 5

natural selection IX, XI, 1, 2, 16, 20, 21, 24, 33, 34, 36, 38–42, 44, 45, 50, 51, 64, 68, 80, 90, 116, 121, 124, 126–129, 132, 133, 135, 140, 141, 143, 144, 150, 151, 153, 156, 163, 165, 169, 170, 171, 173, 175, 176, 178, 184, 188, 192, 197, 198, 200, 202, 205, 207, 210, 216
natural selection for a given environmental pressure may select different features in different taxa 43
neck 11, 12, 72, 99, 106, 121, 143, 149, 156, 202
Neotetrapoda 30, 53, 58, 77, 113, 117, 134, 165, 182, 183, 202, 206, 207
neural arch 13, 207
New Synthesis 24, 193
non-monophyletic groups
  not subject to an evolutionary process 40
Nothosauria 47
notochord 11, 16, 99, 120, 121, 143, 149, 157, 168, 202, 207

Obruchevichthys 71, 103
Old Red Sandstone 25
Onychodontida 73
opercular opening 11
operculum 12
Ossirarus 120
osteichthyans 5, 18, 28, 67, 91, 92, 95, 97, 99, 152, 165, 185, 197, 198
Osteichthyes 5, 7, 16, 18, 25, 65, 66, 73, 90, 94, 122, 138, 139, 158, 182, 184, 186
Osteolepiformes 46, 72, 113
Osteolepis 72

Pachypleurosauria 47

paired locomotor appendages 15, 27, 51, 80, 81, 83, 85, 86, 94, 98, 103, 106, 107, 115, 116, 122, 127, 129, 140–145, 147, 149, 156, 162, 165, 168, 184, 199, 204, 205
Panderichthyidae 50, 72, 88, 100, 107, 113, 143, 158, 203
panderichthyids 27, 50, 102, 200, 203, 205
Panderichthys 50, 72, 103, 144, 203
Pantodon 54
paraphyletic 6, 7, 11, 52, 56, 62, 63, 66, 72, 91, 92, 100, 113, 152, 197, 203, 206
pectoral fin 12, 36, 55, 56, 58, 88, 124, 147, 155, 200
pectoral girdle 11, 12, 16, 84, 99
Pederpes 89, 100, 107, 117, 118, 120, 121, 134, 138, 139, 144, 182
pelvic bone 12
pelvic fins 27, 55, 62, 98, 127, 203
pelvic girdle 15, 55, 88, 103, 122, 143, 145, 156, 168, 184
pelvic or ventral fin 12
perichondral 18
periophthalms 28
Periophthalmus 8, 27, 91, 103, 212
Perittodus 120, 121
Permian 21, 25, 26, 53, 55, 58, 59, 65, 117, 155, 209
Petromyzontoidea 5
Phyllolepis 36, 124
phylogenetic relationships
  shed light on the issue of macroevolution 40
Phylogenetic Systematics xv, 1, 198, 199, 203, 204, 206, 216
Pisces 11, 65, 200, 203, 204
Placodermi 5, 6, 11, 49, 60–62, 64–66, 90
placoderms 48, 52, 59–63, 66, 98, 127, 159, 201, 213
Placodontia 47
plants 23, 24, 26, 60, 171
plesiomorphy 3, 203, 204
Plesiosauria 47
pleural ribs 13, 16
polyphyletic 30, 40, 45, 46, 48, 49, 91, 128, 134, 147, 156, 165, 173, 205
Polypterus 26, 54–56, 92, 93, 125, 147
porolepiform 19
porolepiforms 26, 200, 205
Porolepis 19
practically all characters are exaptations 125
preadaptation
  an exaptation considered before the fact 129
  dealing with morphology 166
  nine categories 169
prospective adaptation 169
Proterogyrinus 118
Protopterus 28, 73, 83, 92
punctuated equilibrium 39, 136

radiation 24, 52, 154, 211
radius/ulna 83
Reptilomorpha 71, 117, 118
return of tetrapods to water 46
returns to an aquatic lifestyle 118
reversions 108
rhipidistians 26, 73, 142
Rhizodontida 71
Romer 9, 26, 27, 31, 54, 65, 72, 79, 89, 90, 103, 205, 215

Sarcopterygii 3, 5, 6, 9, 16–19, 25, 45, 49, 55, 58, 68, 71–73, 81, 82, 85, 104, 108, 110, 111, 122, 127, 128, 138, 140, 144, 147, 154, 156, 182, 184, 186, 207
Science
  crystalized onto sectarian paradigms 194
  permanently revolutionary 194
scorpions 29, 30
selective forces 53
selective pressures x, 24, 26, 27, 30, 36, 45, 48–50, 56–58, 60, 79, 91, 100, 104, 118, 119, 121, 124, 127, 128, 135, 141, 146, 147, 149, 150, 154–156, 183, 197
self-reference of characters 148
semiotic
  genesis of new ideas in science 194
shoulder girdle 19
Silurian 21, 24, 29, 30, 49, 51–53, 58, 64, 65, 138, 139, 145, 157
siluriform
  becoming independently depressiform 48
Simpson 168–175, 177, 179, 181, 193, 215
skeletal struts 82
skull 5, 12, 18, 19, 41, 60, 61, 99, 100, 118, 120, 121, 168, 199, 202, 205

snakelike body form 49, 216
spandrels 39, 42, 130, 205, 211
Spathicephalus 118
subopercle 12, 122, 168
synergy among arguments 152
Systematization 1, 206

Teleostei 7, 15, 16, 36, 46, 48, 54, 56–59, 73
Teleostomi 5
teleosts 7, 8, 11, 15, 19, 27, 28, 48, 55, 89, 91, 94, 97, 103, 115, 126, 143, 150
Temnospondyli 77, 89, 120
temperature 25, 27, 28, 214
terrestriality IX, 5, 11, 16, 20, 21, 26, 31, 32, 49, 52, 54, 56, 62, 64, 66–69, 72, 85, 86, 88, 90–92, 94, 97, 100, 102–108, 111, 114–116, 118–121, 129, 134, 138–144, 146, 148, 149, 152–158, 160, 162, 165, 166, 168, 174, 176–178, 182, 184–187, 207, 209, 216
tests
  of Andrews et al. 186
Testudinata 47
Tetrapoda IX, 1, 3, 6, 9, 11, 20, 30, 32, 45, 50, 51, 53, 55, 57, 58, 60, 71, 72, 77, 88–90, 94, 99, 100, 111, 113, 117, 118, 120, 121, 127, 134, 138, 140, 141, 143, 153–158, 161, 168, 182, 184, 186, 198, 203, 205, 207, 216
tetrapodomorphs 71, 72, 88, 183, 207
tetrapods IX, XI, 18, 20, 25–27, 32, 45, 46, 50, 53, 54, 56, 58, 59, 62, 72, 79, 80–82, 86, 88, 89, 94, 100, 102, 103, 107, 108, 113–115, 117, 118, 120, 121, 123, 138, 139, 141, 142, 149, 154, 165, 185, 197, 201, 202, 205, 206, 209, 211, 214, 215
Thalattosauria 47
tibia/fibula 83
tidal simulations 27
Tiktaalik 50, 72, 88, 89, 100, 103, 113, 143, 145, 203, 215
time needed for
  tinkering 135
tool
  man-made 34, 68, 80, 162
trails 29, 50, 51, 144
Tulerpeton 103, 108

undulatory movement 97
unlikely to be exaptations 126

Ventastega 71, 103
Vertebrae 3, 15
vertebral column 11, 13, 88, 105, 153, 154
Vertebrata X, 3, 111, 211, 216

Wallace IX, 132, 210
way of life 42, 48, 51, 118, 155, 169, 181
Whatcheeria 89, 117, 118, 134
Whatcheeriidae 108, 117, 118, 134, 182
Williams X, 35, 38, 44, 79, 132, 162, 186, 210, 216

zygapophyses 15, 89, 116, 118, 120, 121, 131, 134, 144, 146, 150, 153, 154, 156, 168, 184

www.ingramcontent.com/pod-product-compliance
Lightning Source LLC
Chambersburg PA
CBHW071323230926
51849CB00002B/299